IET ENERGY ENGINEERING SERIES 109

Cooling of Rotating Electrical Machines

Other volumes in this series:

Cooling of Rotating Electrical Machines

Fundamentals, modelling, testing and design

Dave Staton, Eddie Chong,
Steve Pickering and Aldo Boglietti

The Institution of Engineering and Technology

Published by The Institution of Engineering and Technology, London, United Kingdom

The Institution of Engineering and Technology is registered as a Charity in England & Wales (no. 211014) and Scotland (no. SC038698).

The Institution of Engineering and Technology
Futures Place
Kings Way, Stevenage
Hertfordshire, SG1 2UA, United Kingdom

www.theiet.org

British Library Cataloguing in Publication Data
A catalogue record for this product is available from the British Library

ISBN 978-1-78561-351-7
ISBN 978-1-78561-352-4

Typeset in India by MPS Limited
Printed in the UK by CPI Group (UK) Ltd, Croydon

Contents

List of figures

List of tables

About the Authors

Dave Staton (PhD) is the founder and president of Motor Design Ltd (MDL), UK. MDL's Motor-CAD Therm software focuses on electric motor cooling system analysis and design. Previously, David has worked on motor design and in particular development of motor design software at Thorn EMI, the SPEED Laboratory at Glasgow University and at Emerson Electric.

Eddie Chong (PhD) is the technical lead (Asia) of Motor Design Ltd, UK. He has been working on research and consultancy projects on the thermal management of electrical machines over the last 10 years. Also, he has worked on the development of Motor-CAD software for more advanced cooling methods by combining analytical and numerical modelling techniques together with experimental methods for high-power dense electrical machines.

Steve Pickering is a professor of mechanical engineering in the Department of Mechanical, Materials and Manufacturing Engineering, The University of Nottingham, UK. He has been undertaking research on the ventilation and cooling of rotating electrical machines for over 25 years combining both experimental and computational investigations. In recent years, he has used numerical modelling techniques to develop innovative thermal management strategies for high-power density electrical machines for a variety of applications.

Aldo Boglietti is a full professor of Electrical Machines and former head of the Electrical Engineering Department of the Politecnico di Torino, Italy. He is an IEEE fellow member. He was Chair of the Electrical Machine Committee of IEEE IAS (2011–2013), and of IEEE IES (2009–2010). He has written over 270 papers on International Journals and Conferences. In 2020, he was recognized of the ICEM Arthur Ellison Outstanding Achievement Award.

Nomenclature

Roman

A	cross-sectional area (m^2), convective surface area (m^2)
B	flux density (T)
c, c_p	specific heat (J/kg/K)
C	thermal capacitance (J/K)
d, D	diameter (m)
f	friction factor (dimensionless), electric frequency (Hz).
F	force (N)
g	gravitational acceleration (m/s^2).
h	convection heat transfer coefficient (W/m^2/K).
H	pitch-circular radius of rotor ducts (m)
J	rotational Reynolds number (dimensionless)
k	thermal conductivity (W/m/K), turbulent kinetic energy (m^2/s^2)
K	pressure loss coefficient (dimensionless)
l, L	length (m)
$m, \dot{m}$	mass (kg), mass flow rate (kg/s)
Nu	Nusselt number (dimensionless)
p	static pressures (Pa)
P	power (W)
Pr	Prandtl number (dimensionless)
P'	perimeter (m)
q''	heat flux (W/m^2)
q'''	rate of volumetric heat generation (W/m^3)
Q	volumetric flow rate (m^3/s)
q	heat transfer (W)
R	radius, local radius (m)
R	thermal resistance (K/W), flow resistance (kg/m^7), electrical resistance (Ω)
Re	axial flow Reynolds number (dimensionless)

S	gap size (m)
S_{ij}	strain-rate tensor (s^{-1})
t	time (s)
T	temperature (°C or K), torque (Nm)
Ta	Taylor number (dimensionless)
U	axial flow velocity (m/s), overall heat transfer coefficient
V	velocity (m/s)
V_T	tangential velocity (m/s)
W	loss (W/kg)
Z	elevation above a reference plane (m)

Greek

α	thermal diffusivity (m^2/s), temperature coefficient (1/K)
Δ	finite difference
δ_{ij}	Kronecker delta function
ε	turbulent dissipation rate (m^2/s^3)
λ_t	turbulent thermal conductivity (W/m/K)
μ	dynamic viscosity (Pa s)
ν	kinematic viscosity (m^2/s)
ω	specific dissipation rate (s^{-1})
ω_m	angular velocity (rad/s)
Φ	dissipation function
ρ	density (kg/m^3), electrical resistivity (Ωm)
τ	shear stress (Pa), time constant (s)
τ_{turb}	Reynolds stress tensor

Subscripts

h	hydraulic
i	inner, inlet
∞	fully developed flow
m	mean
n	in normal direction
o	stationary, outer, outlet
r	rotational
s	stationary, steady-state
t	turbulence, in tangential direction

Chapter 1

Introduction

Thermal analysis is becoming a more important component of the electric motor and generator design process due to the increasing demands of power output and efficiency for reduced weight and cost. Moreover, the applications for electric machines are increasing dramatically with CO_2 and urban pollution reduction involving developments of cleaner eMobility and renewable energies.

For example, the drive for electrification in the automotive industry has resulted in high-power electric motors becoming one of the key components of the electric powertrain. Increasing power and decreasing mass and space requirements see conventional power density limits being pushed with thermal management of electric motors becoming critical to ensuring reliability and robustness. Electric machine cooling is essential to dissipate the heat generated in the machine while keeping the machine temperature below its thermal limit to avoid insulation breakdown.

In the past, small general-purpose electrical machines tended to have quite simple cooling systems such as the use of natural convection or forced convection by blowing air over the fins outside the motor using fans. Large-sized machines tended to use much more complex cooling systems such as through ventilated cooling with axial and radial cooling channels. For closed-circuit cooling, a heat exchanger is critical to reduce the coolant temperature to a safe level before it re-enters into the machine. The coolant inside the machine could be air or hydrogen depending on machine requirements. Medium-sized machines would typically have blown over-cooling or through ventilated cooling using axial channels. With the demands of higher power density electric machines, as required by new applications such as servo and eMobility, there is a trend to use more effective cooling systems such as water jackets and oil spray cooling.

In the future development of electrical machines, one of the solutions is higher speed to meet the power requirements with limited weight and space. To cool an electrical machine effectively, one needs to have a detailed understanding of the losses and their distribution. For high-speed machines, this leads to increased frequency-related losses in the magnetic components such as eddy current loss and hysteresis loss. High-speed operations can also lead to substantial AC loss in the conductors due to proximity and skin effects. In particular, the proximity loss is more concentrated in the winding area at the slot opening during high-speed operation. This gives much more severe winding hotspot temperatures when compared to the thermal model without proximity loss being considered. Therefore, it is essential for machine designers to combine both electromagnetic and thermal analysis to calculate the losses and to determine how the losses may be extracted in the cooling system.

Furthermore, thermal management of electrical machines tends to make use of more advanced materials. When compared to conventional electrical steels, new magnetic materials give lower losses under the same operating condition. Rare earth (samarium cobalt and neodymium) magnets are becoming common for permanent magnet machines because of their higher magnetic properties. Neodymium magnets give the highest magnetic properties but their flux density reduces considerably with increased temperature and they have lower maximum operating temperatures. In contrast, samarium cobalt magnets have low magnetic degradation at elevated temperatures but are very expensive. To meet the increasing demand, machine designers need to optimize not only the electrical design but also need to have an efficient cooling system. Conventional insulation materials have low thermal conductivity and result in high-temperature gradients in the machines. More thermally conductive insulation materials such as slot liners, wire enamel coatings, and impregnation/potting materials are being developed to improve heat extraction. More effective convective cooling such as oil spray or oil-flooded cooling becomes an attractive solution because it allows direct contact between coolant and heat sources.

Currently, eMobility is moving at a great pace. In the past, there has been much development in railway electrification. Automotive electrification is currently developing fast and the future trend is in aerospace electrification. This leads to a larger challenge for high power density electrical machines as minimum system weight (the electrical machine, the power electronics drive, and their cooling systems) is even more important in aerospace applications. With a high-voltage system, the machines are also subjected to the detrimental effect due to partial discharge. Moreover, working at a high altitude with reduced atmospheric pressure gives more complications. Consequently, aerospace applications where electrical machines are used for propulsion motors and generators in hybrid systems give some of the most demanding thermal management applications for conventional electrical machines. In the future, super-conducting electrical machines may be used as these offer the potential for high-power densities; however, the cryogenic cooling systems required are beyond the scope of this book.

This book is aimed at electrical engineers who are familiar with electrical machine design but may be less familiar with the concepts in heat transfer and fluid flow. Chapter 2 gives an introduction to the principles of heat transfer that commonly apply to electrical machines. This includes the principles of conduction, convective, and radiation heat transfer, and a number of more specialized topics in heat transfer, such as finned surfaces, heat exchangers, and other topics that are related to electrical machines.

Chapter 3 gives an introduction to the principles of fluid flow. This starts with some of the basic concepts in fluid mechanics and then describes in more detail fluid flow calculations of particular relevance to electrical machines, including flow in ducts, in which details are given for calculating losses; the performance of fans; flows in annular gaps; and flows in rotating ducts and over-rotating disks.

Chapter 4 describes the thermal modelling of electrical machines. A detailed description of the lumped-parameter thermal network (LPTN) method is given that includes details of how elements of electrical machines can be represented using this technique. This is followed by examples of the application of the LPTN modelling method to a range of different electrical machines employing different cooling techniques ranging from totally enclosed fan-cooled to closed-circuit internally cooled machines as well as various applications of liquid cooling.

Chapter 5 describes more advanced computational methods for modelling fluid flow and heat transfer within electrical machines. This includes the use of finite-element analysis (FEA) and computational fluid dynamics (CFD). The computational methods are outlined and the application to electrical machines is described giving examples.

Chapter 6 describes test methods for electrical machines. This includes a description of experimental techniques that may be used for taking both thermal and flow measurements. Methods for determining losses are given as well as methods for calibrating thermal models.

Finally, Chapter 7 gives a number of case studies of the application of design tools to the thermal design of electrical machines.

Chapter 2

Fundamentals of heat transfer

Heat transfer deals with the rate of heat flow as a result of temperature differences. There are three principal mechanisms: conduction, convection, and radiation, and all are relevant to electrical machines. Conduction occurs principally in solids and the rate of heat flow is determined by the temperature gradient and the thermal conductivity of the material. In electrical machines, it is of concern in dissipating heat in solid regions, such as the electrical conductors, magnetic iron, insulating materials, and frame.

Convection is the dominant form of heat transfer in liquids and gases and is associated with the transfer of heat by movement of the fluid. Air or liquid cooling systems are used on most electrical machines and the nature of the fluid flow, in terms of the type of fluid and flow pattern over the surfaces being cooled, determines the rate of heat transfer by convection.

Radiation heat transfer is in the form of electromagnetic radiation and in electrical machines occurs principally between solid surfaces separated by air gaps. Solids and liquids used in electrical machines can be considered to be opaque to thermal radiation and air can be considered to be transparent and not interact with radiation heat transfer.

This chapter gives a basic introduction to heat transfer and focuses on aspects that are particularly relevant to electrical machines. Many textbooks give more detailed coverage of the subject that may be referred to and examples are given in Refs [1–5].

2.1 Conduction heat transfer

2.1.1 Fourier's law

Conduction takes place in any substance in which there is a temperature gradient and in opaque solids, it is the only mechanism of heat transfer. Heat energy or internal energy can be understood on a molecular scale as kinetic energy in the form of vibrating atoms. In materials at high temperatures, the atoms have more kinetic energy and are vibrating fast. Atoms are continuously colliding with one another and exchanging kinetic energy. In a region of a temperature gradient, there will thus be the tendency for the energy to be diffused as the kinetic energy is distributed more evenly through the material. Consequently, heat flows from hot to cold through this redistribution or diffusion of the internal energy. The rate at which heat diffuses or conducts through a material depends linearly on the temperature gradient and the physical property known as thermal conductivity. This is described by Fourier's law in the following equation:

$$q_x = -kA\frac{\partial T}{\partial x} \tag{2.1}$$

It may be expressed more generally as a heat flux (heat flow per unit area):

$$q_x'' = -k\frac{\partial T}{\partial x} \tag{2.2}$$

2.1.2   Thermal conductivity

The rate of heat conduction depends upon the thermal conductivity (k) of the material. In non-metallic substances, heat conduction takes place by the transfer of kinetic energy between atoms, as described above, and this is known as *lattice vibration*. In metals, heat internal energy is also transferred by the movement of free electrons in between the atoms (known as *electron flow*) and this significantly enhances the rate of heat transfer. So, metals, in general, have higher thermal conductivities than non-metals and typical values are given in Table 2.1. The similarity in the mechanisms for heat conduction and electrical conduction in metals means that, in general, good electrical conductors are also good heat conductors. The notable exception to this is diamond, which has the highest thermal conductivity of any natural substance yet, is an electrical insulator.

2.1.2.1   Variation in thermal conductivity

Although thermal conductivity is often referred to as a constant in Fourier's law, thermal conductivities are not constant but vary with temperature and, in some substances, with the direction of heat flow. In the temperature ranges associated with electrical machines, these variations of thermal conductivity with temperature are typically not very large. The thermal conductivity of copper decreases by 3% as temperature increases from 0 °C to 200 °C and this will only have a small effect on the thermal performance of an electrical machine. Given the uncertainties in other quantities, such as thermal contact resistance and heat transfer coefficient, it would not normally be necessary to account for the variation of thermal conductivity with temperature in most calculations. A larger effect is caused by the corresponding increase in electrical resistance which does significantly increase the losses.

2.1.2.2   Proprietary insulations

It should be noted that the thermal conductivities shown in Table 2.1 are indicative values, particularly for polymer insulating materials. Generally, insulating materials have significantly lower thermal conductivities than metals and can add significant thermal resistance when used as electrical insulation. Material suppliers are continuing to develop proprietary insulations, for example, for winding insulation, slot liners, and impregnation resins with improved thermal and electrical properties.

Table 2.1   Thermal conductivity of some typical materials used in electrical machines from Bejan and Kraus [5], Motor-CAD [6], and Bolton [7]

Material	Thermal conductivity (W/m/K)
Gases	
Air (20 °C)	0.026
(100 °C)	0.032
Hydrogen (20 °C)	0.18
Argon (20 °C)	0.018
Resins and plastics (20 °C)	
Polytetrafluoroethylene (PTFE)	0.23
Perspex (acrylic)	0.2
Phenolic resin	0.23
Polyether ether ketone (PEEK)	0.25
Epoxy resin	0.22
Nomex (25–150 °C)	0.10–0.14
Iron/steel (20 °C)	
Low carbon steel (<0.2% C)	50
Stainless steel (types 304/316)	15
Pure iron	81
Cast iron (5% C)	52
Iron 1% silicon	42
Iron 2% silicon	28
Iron 5% silicon	19
Aluminium alloys (20 °C)	
Aluminium (pure)	237
1,000 series (unalloyed)	220–230
2,000 series (copper alloys)	120–160
5,005 Electrical conductors	201
Copper pure (20 °C)	401
High copper alloys	340–370

The thermal conductivity of polymers can be increased, for instance, by including powdered non-metallic fillers with higher thermal conductivity. For example, alumina has a thermal conductivity up to 30 W/mK, depending on purity, and modest proportions of filler can significantly increase the thermal conductivity of a polymer. Increases in thermal conductivity of electrical insulation are to be desired and can result in significant improvement in the performance of machines. Examples of the effect of insulation thermal conductivity on electrical machine performance are given in Chapter 7, Section 7.1. Machine designers are therefore advised to be aware of the insulation materials available.

2.1.2.3   Variation of thermal conductivity with direction

Thermal conductivity varies with direction in many materials that possess an anisotropic structure. In electrical machines, anisotropic properties result from the way in which materials are used within the machine rather than from any inherent characteristic of the materials used. The use of laminations to form magnetic cores results in a structure that has a reduced thermal conductivity perpendicular to the laminations due to the thermal contact resistance in between each lamination and the surface varnish. Contact resistance is discussed in more detail later in this chapter. Similarly, in the build-up of coils, the layers of relatively low thermal conductivity insulation between the copper conductors result in an effective thermal conductivity perpendicular to the conductors that may be one or two orders of magnitude lower than the thermal conductivity along the conductors.

2.1.3   Thermal conduction in one dimension at steady state

2.1.3.1   Flat walls

Heat conduction across a flat wall of material may be considered to be in one dimension if the only temperature gradient is through the thickness (Figure 2.1). At steady state, Fourier's law can be integrated to give the following formula:

$$q_x = \frac{kA(T_1 - T_2)}{\Delta x} \tag{2.3}$$

This may be expressed in terms of the heat flux as:

$$q_x^{''} = \frac{kA(T_1 - T_2)}{\Delta x} \tag{2.4}$$

The similarity with electrical conduction is recognized and heat conduction may be expressed in terms of thermal resistance as:

$$q_x = \frac{(T_1 - T_2)}{R_{th}} \tag{2.5}$$

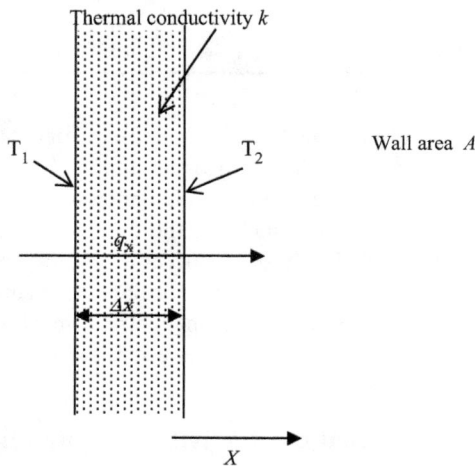

Figure 2.1   Heat conduction through a flat wall

where

$$R_{th} = \frac{\Delta x}{kA} \tag{2.6}$$

The heat conduction is equivalent to current in an electrical circuit and the temperature difference is equivalent to the driving voltage potential. Heat transfer can therefore be represented diagrammatically as a resistance element as shown in Figure 2.2:

The resistance network method is used extensively in the analysis of the thermal performance of electrical machines and will be described in more detail in later sections of this book.

Where the wall through which the heat is conducted is made up of layers of material, the heat flow can be analyzed as thermal resistances in series (Figure 2.3). In many situations, heat is transferred through the wall from fluids on either side and the thermal resistance due to convective heat transfer needs to be added. Convection heat transfer from a hot surface to a fluid is expressed as:

$$q = hA\left(T_s - T_f\right) \tag{2.7}$$

Rewriting this equation in terms of a thermal resistance

$$q = \frac{T_s - T_f}{R_{th}} \tag{2.8}$$

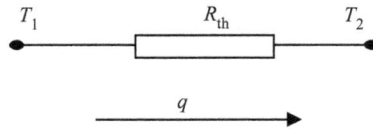

Figure 2.2   *Representation of a thermal resistance*

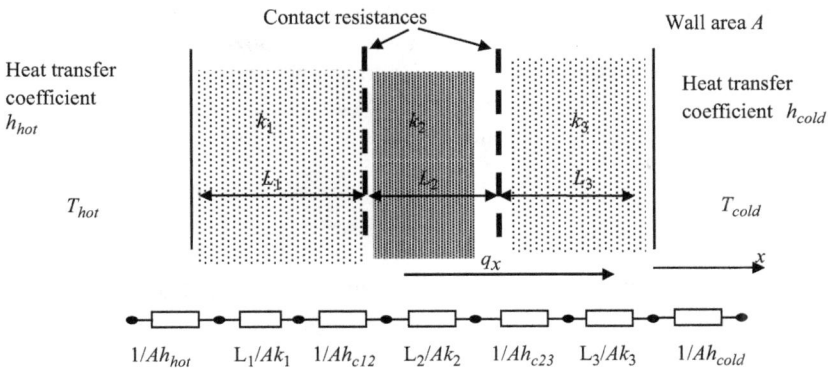

Figure 2.3   *Heat conduction through a composite wall*

results in the following expression for thermal resistance due to convection at a surface:

$$R_{th} = \frac{1}{hA} \tag{2.9}$$

There may also be thermal contact resistances between the layers of material in the wall that must be accounted for, as described in Section 2.1.5.

Note that thermal resistances usually have the units K/W, but may also be expressed per unit area, particularly if heat flux is to be calculated. The thermal resistance of a layer of solid material of thickness Δx, would then be given as $R_{th} = \Delta x/k$ Km²/W.

As the heat flows through each thermal resistance, in turn, the total thermal resistance can be calculated by summing the individual thermal resistances:

$$R_{tot} = \frac{1}{A} \left[\frac{1}{h_{hot}} + \frac{L_1}{k_1} + \frac{1}{h_{c12}} + \frac{L_2}{k_2} + \frac{1}{h_{e23}} + \frac{L_3}{k_3} + \frac{1}{h_{cold}} \right] \tag{2.10}$$

So, the heat flow is given by:

$$q_x = \frac{T_{hot} - T_{cold}}{R_{tot}} \tag{2.11}$$

Alternatively, the heat flow can be described in terms of an *overall heat transfer coefficient* or *U-value*. This is given as:

$$q_x = UA(T_{hot} - T_{cold}) \tag{2.12}$$

where the *U-value* is the reciprocal of the total thermal resistance per unit area and is given by:

$$U = \frac{1}{\left[\frac{1}{h_{hot}} + \frac{L_1}{k_1} + \frac{1}{h_{c12}} + \frac{L_2}{k_2} + \frac{1}{h_{e23}} + \frac{L_3}{k_3} + \frac{1}{h_{cold}} \right]} \tag{2.13}$$

2.1.3.2   Cylinders

One-dimensional heat conduction can also occur when the geometry is cylindrical and the only temperature gradient is in the radial direction. Heat conduction through the rotor and stator cores of electrical machines can be analyzed in this way if axial or circumferential temperature gradients are ignored as shown in Figure 2.4.

Fourier's law can be integrated across the cylinder to give the heat flow as:

$$q_r = \frac{2\pi k L}{\ln\left(\frac{r_o}{r_i}\right)}(T_o - T_i) \tag{2.14}$$

The thermal resistance for heat flow through the cylinder is thus given by:

$$R_{th} = \frac{\ln\left(\frac{r_o}{r_i}\right)}{2\pi k L} \tag{2.15}$$

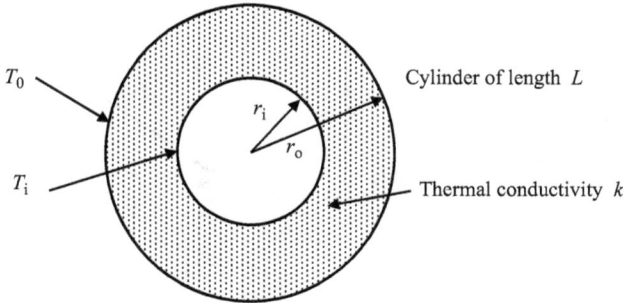

Figure 2.4   Heat conduction through a cylinder

An important feature of this geometry is that, for uniform radial heat conduction, the cross-sectional area through which the heat flows increases as the radius increases. This means that the heat flux reduces as the radius increases. The expression for the thermal resistance due to convection, and also thermal contact resistance, takes the form $\frac{1}{hA}$, where area A is the surface area at the convection boundary or interface between two cylindrical layers ($A = 2\pi r L$). The magnitude of the convection or contact resistance thus depends on the radius at which it occurs.

A composite cylinder comprising layers of different material and thicknesses can be analyzed in the same way as a flat wall using the appropriate thermal resistance terms for radial heat conduction, convection, and thermal contact resistance.

2.1.4   General conduction equation

A general equation can be derived for heat conduction in three dimensions accounting for the heat flows as illustrated in Figure 2.5. This includes heat generation within the volume (q''') (e.g. due to an electric current) and heat storage within the volume, resulting in a temperature change with time.

$$\begin{pmatrix} \text{The net rate of heat transfer} \\ \text{into control volume} \\ \text{by conduction} \end{pmatrix} + \begin{pmatrix} \text{rate of} \\ \text{internal heat} \\ \text{generation} \end{pmatrix} = \begin{pmatrix} \text{rate of} \\ \text{heat} \\ \text{accumulation} \end{pmatrix}$$

Figure 2.5 shows a control volume in Cartesian coordinates, illustrating heat conduction in the three coordinate directions as well as internal heat generation.

The general equation can be derived by applying the energy balance, shown in text equation above, to the control volume and the detailed derivation is given in most textbooks: Simonson [1], Holman [2], Bergman *et al.* [3], and Bejan [4]. Note

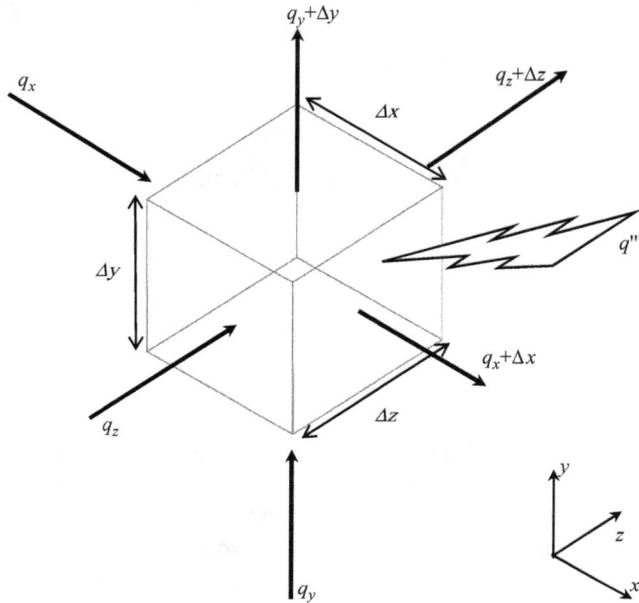

Figure 2.5   A control volume for conduction showing heat flows in three dimensions

that it has been assumed that the thermal conductivity k is the same in all three coordinate directions:

$$k\left[\frac{\partial^2 T}{\partial x^2} + \frac{\partial^2 T}{\partial y^2} + \frac{\partial^2 T}{\partial z^2}\right] + q''' = \rho c \frac{\partial T}{\partial t} \tag{2.16}$$

It is common to divide (2.16) by thermal conductivity to give:

$$\left[\frac{\partial^2 T}{\partial x^2} + \frac{\partial^2 T}{\partial y^2} + \frac{\partial^2 T}{\partial z^2}\right] + \frac{q'''}{k} = \frac{1}{\alpha} \frac{\partial T}{\partial t} \tag{2.17}$$

where $\alpha = k/\rho c$ is known as thermal diffusivity.

Equation (2.17) is the general heat conduction equation in Cartesian coordinates and may be applied to any heat conduction problem, assuming that thermal conductivity is a constant and does not vary with temperature, time, position, or direction.

2.1.4.1   Consideration of laminated materials

In modelling heat conduction in laminated cores of electrical machines, it is not usually computationally efficient to model each lamination individually with thermal contact resistances in between. The effect of the laminations is to significantly decrease the thermal conductivity across the laminations and this can easily be modelled by assuming the laminated core is a homogenous material but with an anisotropic thermal conductivity in which the conductivity along the laminations is that of the lamination material, but the conductivity across the laminations is much smaller, taking into account the thermal contact resistance between the laminations. Typically, the thermal conductivity across the laminations may be less than 10% of the thermal conductivity along the laminations. This anisotropic effect can be accounted for in the general heat conduction equation by the use of different thermal conductivities in each coordinate direction, assuming that the coordinates are aligned with the laminations. Equation (2.17) thus becomes:

$$\left[k_x \frac{\partial^2 T}{\partial x^2} + k_y \frac{\partial^2 T}{\partial y^2} + k_z \frac{\partial^2 T}{\partial z^2} \right] + q''' = \rho c \frac{\partial T}{\partial t} \tag{2.18}$$

where k_x, k_y, and k_z are the thermal conductivities in each coordinate direction.

2.1.4.2 Cylindrical geometries

Equation (2.17) was derived for a rectangular or Cartesian coordinate system. In most electrical machines, the topography is cylindrical and it makes more sense to derive the general conduction equation in cylindrical coordinates where the principal directions are radial, circumferential, and axial.

Expressed in cylindrical coordinates, the general heat conduction equation now becomes:

$$\left[\frac{1}{r} \frac{\partial}{\partial r} \left(r \frac{\partial T}{\partial r} \right) + \frac{1}{r^2} \frac{\partial^2 T}{\partial \theta^2} + \frac{\partial^2 T}{\partial z^2} \right] + \frac{q'''}{k} = \frac{1}{\alpha} \frac{\partial T}{\partial t} \tag{2.19}$$

2.1.5 Thermal contact resistance

When two solid surfaces are in contact with each other, the conduction of heat across the boundary between the surfaces is impeded by the nature of the surfaces making contact and a thermal resistance occurs. This is known as thermal contact resistance. It arises as solid surfaces are not completely flat and smooth and on a microscopic level, contact is only made over a proportion of the total area in contact, as illustrated in Figure 2.6. The gaps in between the contacts are usually filled

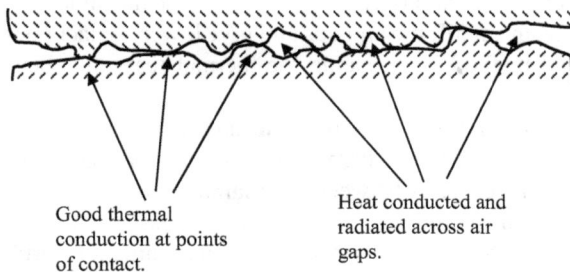

Good thermal conduction at points of contact.

Heat conducted and radiated across air gaps.

Figure 2.6   Nature of contact between solid surfaces—an illustration of a contact on a microscopic scale

with air and heat will be conducted through the air and across the direct points of contact, heat will also be radiated across the air gap. Examples of areas in electrical machines where it is of importance are in the conduction of heat between laminations in the magnetic cores, in the contact between the stator core and frame of a machine, and the contact between coils and laminations in stator slots.

The contact resistance is often expressed as a heat transfer coefficient (h_c) [W/ m^2 K] or a thermal resistance (R_c) [K m^2/W]. In electrical machines, it is also often expressed as an equivalent air gap thickness (t). These three expressions are related as follows:

$$R_c = \frac{1}{h_c} = \frac{t}{k} \qquad (2.20)$$

where k is the thermal conductivity for air. Taking a value for k at 25 °C of 0.026 W/m K, an equivalent air gap thickness t expressed in mm is given by:

$$t = 26 R_c [\text{mm}] \text{ or } t = \frac{26}{h_c} [\text{mm}] \qquad (2.21)$$

Expressing the thermal contact resistance in terms of an equivalent gap is a convenient way of representing the contact, but it implies that conduction through the medium in the gap is the mode of heat transfer. This is not always the case and where there is air present in the gap, or another fluid that is transparent to thermal radiation, then heat transfer also takes place by radiation. Nevertheless, the representation of the thermal contact resistance as an equivalent air gap is an appropriate means of expressing the resistance.

An example of the way that it can be included in thermal calculations is illustrated in Figure 2.3 and (2.10).

The contact resistance depends upon the nature of the surfaces in contact. Flat smooth surfaces have lower contact resistances as a greater proportion of the solid material is in contact. High contact pressures encourage lower thermal resistance as the solids deform to give a higher contact area. Lower thermal contact resistances can be achieved when softer materials such as aluminium are used rather than steels. Lower thermal resistances can also be achieved by filling the gaps with a substance with a higher thermal conductivity than air. When a machine has been vacuum impregnated, the resin can fill the air gaps and give lower contact resistances. This is a particularly good method for reducing the thermal resistance between the stator coils and stator core in many machines.

Some typical values of effective thermal interface gap are given in Table 2.2.

For laminations, the effective thermal conductivity through the lamination stack involves many thermal contacts between each lamination and is a complex function of such aspects given as follows: the clamping pressure; lamination thickness; stacking factor; lamination surface finish; and interlamination insulation material. Typical ratios of radial-to-axial thermal conductivity are 20–40 [6]. Further discussion of thermal contact resistance and the effective thermal conductivity of laminations is given in Chapter 4, Section 4.1.5, in which further data is given.

Table 2.2 Some representative values of thermal contact resistance

Contact	Thermal resistance (K m²/kW)	Equivalent air gap (mm)
Stator to housing contact resistance measured for a range of standard totally enclosed fan-cooled (TEFC) induction machines from frame size 112–500 frame with cast iron and aluminium frames. Staton *et al.* [8]	0.38–3.0	0.01–0.077
Machined stator with water-cooled frame and shrink fit pressure 13, 21, and 39 MPa. Kulkani *et al.* [9]	0.141, 0.133, and 0.098	0.0037, 0.0035, and 0.0025
Machined and non-machined stator with a water-cooled frame, thermal paste, and shrink fit pressure of 17 MPa. Kulkani *et al.* [9]	0.112, 0.151	0.0029, 0.0039

In electronic circuits, components that dissipate significant quantities of heat are often attached to heat sinks with a heat conductive paste between the surfaces in contact to reduce the thermal contact resistance. When heat fluxes are high, soldered joints provide a very low thermal resistance, though permanent, interface. A range of proprietary thermal interface materials is available; some based on soft metallic foils or others utilizing low melting temperature metals. While these interfaces may be cost-effective for relatively small areas in electronic circuits, they are generally not viable in electrical machines.

More information on thermal contact resistance is also given in Bejan and Kraus [5].

2.1.6 *Boundary conditions*

In solving the heat conduction equation (2.16), it is the boundary conditions that uniquely define the solution to a particular problem. It is thus important to correctly define boundary conditions and this section explains the typical boundary conditions that are found in heat conduction problems. A knowledge of the types of boundary conditions and their significance is particularly important when using advanced numerical techniques such as finite-element analysis (FEA) and computational fluid dynamics (CFD) software packages. These packages are designed to relieve the burden of solving the governing equations and do so accurately and robustly. However, the results will be inaccurate if the boundary conditions have not been correctly identified and specified.

2.1.6.1 Defined temperature boundary condition

Specifying the temperature at a particular surface is the easiest boundary condition to apply. But it is only rarely that the temperature of a surface will be known in practice. It is more usual for the temperature of a fluid to be specified (for instance, the temperature of the ambient air at inlet to the fan in a TEFC motor) and then a convective boundary condition at a surface is most appropriate.

2.1.6.2 Defined heat flux boundary condition

If the heat flow is known at a boundary, then it is possible to define the temperature gradient at the boundary from the expression for Fourier's law given earlier:

$$q_x^{''} = -k\frac{\partial T}{\partial x} \tag{2.22}$$

In electrical machines, the heat input often comes from the electrical losses due to ohmic heating in conductors or electromagnetic losses in the magnetic cores. In these cases, the heat flux at a boundary may be known and hence the temperature gradient. However, in most cases, although the total heat flow from a conductor or a core may be known, the distribution of the heat flux over the surface may not and so the heat flux may vary in an unknown way. For instance, consider the rotor of an induction motor. If heat conduction along the shaft is insignificant, then all the rotor losses must leave the surface of the rotor. However, that is not to say that the heat flux will be uniform over the rotor surface and it will depend upon the distribution of the losses within the rotor and the variation of convective cooling over the surface.

There are two special cases where the heat flux, and hence the temperature gradient, is zero at a boundary. The first is the case of an *insulated boundary*. If the insulation is infinitely thick then there will be no heat flow. In practice, through surfaces that are well insulated thermally, there will generally be an insignificant heat flow and the zero heat flux, or zero temperature gradient may apply. The second special case is that of a plane of *thermal symmetry* where, by definition, there is no heat flux at the boundary. In electrical machines, it is often practical to assume that rotors and stators are rotationally symmetric and so thermal symmetry may be assumed on radial planes at the centre of the conductors or the teeth in the iron core.

2.1.6.3 Convective boundary condition

This is by far the most common boundary condition since most surfaces are exposed to a fluid where convection will take place. The convection boundary condition is based on an energy balance at the surface where it is recognized that the heat flux at the surface of the solid due to conduction is equal to the heat flux due to convection into the fluid. So, at the surface:

$$-k\frac{\partial T}{\partial x}\bigg|_{wall} = h\left(T_{wall} - T_{fluid}\right) \tag{2.23}$$

where x is in the direction perpendicular to the wall.

2.1.7 Solution of steady-state heat conduction problems in two and three dimensions

In problems where the heat flow is only, or predominantly, in one direction, it is usually possible to solve a heat conduction problem by assuming one-dimensional heat flow as described in Section 2.1.3 above.

However, when the configuration is such that the heat flow is in more than one dimension, the analysis is usually much more complicated, and more elaborate solution methods are required. Examples in electric machines might be the heat conduction in the stator windings—where heat is conducted along the windings from the end windings and conducted from the sides of the coils into the stator teeth and from the top of the coil directly into the air gap. Analysis of such a three-dimensional heat flow is complicated further by the inhomogeneous nature of the winding with layers of insulation of various thicknesses surrounding the copper conductors. In electric machines, although the geometries of many of the components are straightforward, for instance, the stator coils and stator teeth may be approximated as long rectangular components, the construction often makes the conduction inhomogeneous (stator laminations and the insulated conductors in the coils) and the boundary conditions are often complex. For instance, on the stator coils of a large machine with radially ventilated stator where the sides of the coils are partly embedded in the stator slots and partly exposed to the air in the stator ducts. In these situations, the analytical procedures for calculating heat conduction are inadequate for giving detailed temperatures and numerical methods are best used. With the increasing power now available on desktop machines, very complex problems can now be solved accurately in time periods that are practical for industrial design. Many different numerical techniques may be used and these are described in later chapters of this book.

2.1.8   Transient heat conduction

2.1.8.1   Transient conduction phenomena

Transient conduction occurs whenever temperatures are not steady but vary with time. A change in temperature means that the internal heat energy within the object has changed. An increase in temperature is effectively a sink for thermal energy and heat energy is released when temperature decreases. In electrical machines, heat is generated by the machine's losses and these have to be dissipated. Heat dissipation can only occur if there is a driving temperature difference between the machine and its surroundings, or a cooling medium. So, when a machine starts up from cold, the losses are absorbed by the machine structure and it increases in temperature until a point where the rate of heat dissipation due to the temperature difference equals the rate at which heat is generated from the losses.

When an object, initially at a uniform temperature, is subject to a temperature change at its boundary (a hot object being subject to colder surroundings for instance), heat transfer takes place at the boundary where the temperature difference occurs. Then, as temperature gradients develop within the object, heat is conducted towards the boundary and the temperature inside the object starts to fall progressively. Regions toward the centre of the object may not "feel" the effects of the heat transfer at the boundary until sometime after the temperature change is imposed. After a period of time, the temperature within the object will again be uniform. The temperature profiles within the object may be as shown in Figure 2.7.

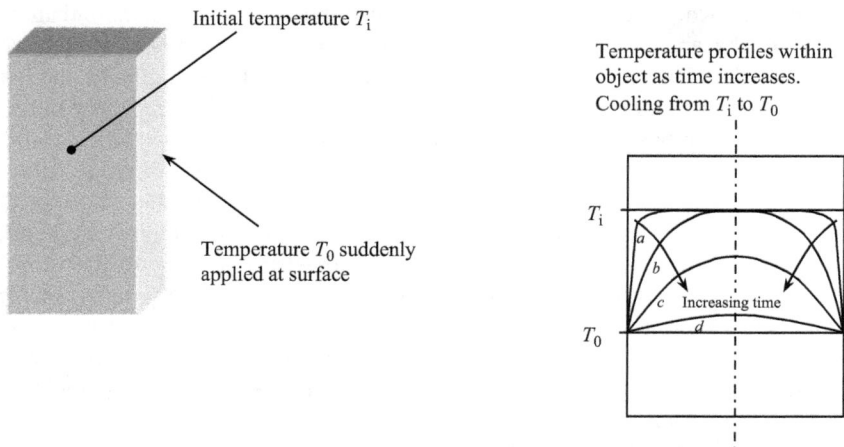

Initial temperature T_i

Temperature T_0 suddenly
applied at surface

Temperature profiles within
object as time increases.
Cooling from T_i to T_0

Figure 2.7   Representation of temperature profiles during transient conduction

The stages in the development of the transient temperature change within an object are illustrated in Figure 2.7. During the *early* time (profile *a*), the temperature change is only apparent near the surface of the object. During the *late* time (profile *d*), the object is again at a near-uniform temperature. These stages can be characterized by a dimensionless number called *Fourier number* (*Fo*). It is defined as follows:

$$Fo = \frac{\alpha\, t}{L^2} \tag{2.24}$$

where
α is the thermal diffusivity of the material defined as $k/\rho c$; t is the time since the start of the temperature transient; L is a characteristic length for the object over which the heat transfer is taking place (e.g. half the thickness for a wall of material).

For a $Fo \ll 1$, a temperature change will only have had an effect near the surface of an object (e.g. profile *a* in Figure 2.7).

For a $Fo \approx 1$, a temperature change will have had an effect through an object (e.g. curve *c* in Figure 2.7).

For a $Fo \gg 1$, the effects of the transient will have spread throughout the object and the temperature changes will be smoothed out and approach a steady state (e.g. curve *d* in Figure 2.7).

In considering transient heat conduction, it is useful to calculate a *Fourier* number to assess how far the transient effects may have penetrated through the object. If the time interval is such that $Fo \ll 1$, then the transient effect will not have penetrated through an object and only the surface will have experienced a temperature change. But if $Fo \gg 1$, then the effects of the transient will have diffused throughout the object and the temperatures will be approaching a uniform steady state.

Example: Consider the thick cylinder shown below. If a heat flux is applied to the inside surface for 1 min, will the outside of the cylinder have changed significantly in temperature? The cylinder is 100 mm thick and made from steel. The characteristic length can be assumed to be 100 mm, the distance that heat is conducted from the inside to the outside of the cylinder:

$$Fo = \frac{kt}{\rho c L^2}$$

For a time of 60 s, $Fo = 0.08$.

This is much less than 1 and therefore after 1 min the effect of the heat flux will not have affected the outside surface of the cylinder and it will take 12 min before the heat has penetrated through the cylinder and much longer before a uniform temperature is reached.

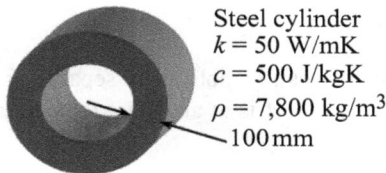

Steel cylinder
$k = 50$ W/mK
$c = 500$ J/kgK
$\rho = 7,800$ kg/m^3
100 mm

2.1.8.2 Solution of transient heat conduction by a lumped capacity method

A simple approach is to use what is known as the lumped capacity method, in which the temperature within in an object, or a particular region of an object, is assumed to be uniform and any temperature gradients are small compared to the temperature difference between the object and its surroundings. Heat exchange with the surroundings is considered and an expression derived to describe how the temperature of the object varies with time.

Consider a hot object in a cooler surrounding as shown in Figure 2.8.

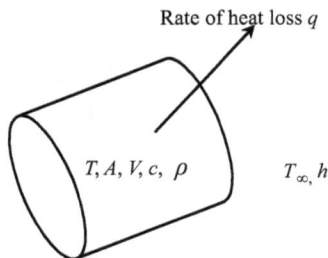

Rate of heat loss q

T, A, V, c, ρ T_∞, h

Figure 2.8   Representation of heat transfer from a solid to a fluid

where A is the exposed surface area of the object, V is the volume.

The rate of heat transfer from the object at temperature T to the surroundings at T_∞, is given by:

$$q = hA(T - T_\infty) \tag{2.25}$$

As the heat is transferred from the object, it will cool down and the rate of cooling is given by:

$$q = -\rho V c \frac{\partial T}{\partial t} \tag{2.26}$$

The minus sign in (2.26) indicates that as heat is lost, the temperature decreases with time.

Combining equations (2.25) and (2.26) gives:

$$hA(T - T_\infty) = -\rho V c \frac{\partial T}{\partial t} \tag{2.27}$$

This differential equation can be solved by separating the variables and integrating them. Given that the object is at an initial temperature of T_0 at time $t = 0$, the result is:

$$T - T_\infty = (T_o - T_\infty)e^{-\left(\frac{hA}{\rho V c}\right)t} \tag{2.28}$$

This equation gives the characteristic exponential cooling curve and describes the well-known phenomenon that an object will cool until it is equal in temperature to its surroundings, i.e., as $t \rightarrow \infty$, so $T \rightarrow T_\infty$.

In (2.28), the term $\rho V c / hA$ is known as the *time constant* as it has the units of time. It indicates how long it will take for an object to change in temperature. It can be shown from (2.28) that the *time constant* indicates how long it takes for the temperature difference to decay to 36.8% of the initial value.

The lumped capacity method also has an electrical resistor–capacitor (RC) circuit analogy as shown in Figure 2.9.

The thermal capacity of the object is represented by the capacitor and the resistance represents the thermal resistance to heat transfer from the object to the surroundings.

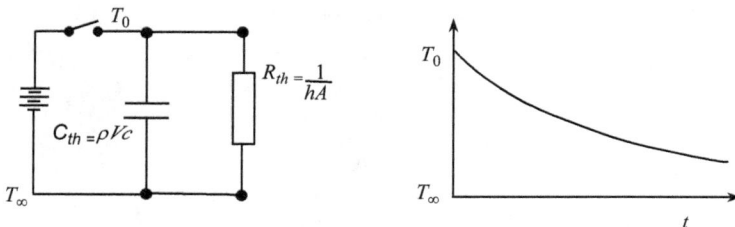

Figure 2.9   Analogy between lumped capacity thermal analysis and an electrical RC circuit

The system is charged up to temperature T_0 by closing the switch. On opening the switch, the charge stored in the capacitor discharges to the ambient potential through the resistance, and the voltage T_0 decreases to T_∞.

2.1.8.3   Applicability of the lumped capacity method

The lumped capacity method assumes that temperature gradients within the object are small compared to the temperature difference between the object and its surroundings. This means that the surface temperature, which determines the rate of heat loss, is the same as the (uniform) temperature within the object, which is used in determining the rate of decrease of internal energy. This implies firstly that the *Fourier number* is large, in that the temperature changes experienced at the surface will have penetrated throughout the object in the time intervals concerned. The internal temperature gradients within an object depend upon the internal thermal conduction resistance, and the temperature difference between the object and its surroundings is determined by the external thermal resistance. The thermal conduction resistance for a flat wall is given by (2.6) and so for an arbitrarily shaped object, the internal thermal resistance ($R_{th\ int}$) can be defined as:

$$R_{th\ int} = \frac{s}{kA} \tag{2.29}$$

where s is a characteristic thickness of the object.

The external thermal resistance ($R_{th\ ext}$) can be defined as:

$$R_{th\ ext} = \frac{1}{hA} \tag{2.30}$$

So, the ratio of the internal to external thermal resistance is found by dividing (2.29) by (2.30). This relationship is non-dimensional and is known as the *Biot number* (*Bi*).

$$Bi = \frac{hs}{k} \tag{2.31}$$

Provided that the *Biot number* is suitably small, the lumped capacity analysis can be considered to be valid. It is generally accepted that when $Bi < 0.1$, the lumped capacity method can give suitably accurate answers.

In summary, the lumped capacity method can be applied when $Bi < 0.1$ and $Fo \gg 1$.

2.1.8.4   Analytical solution of the heat conduction equation

In a case where the $Bi > 0.1$, the lumped capacity method is not appropriate and so the temperature gradients within the object must be considered and the heat conduction equation must be solved. The general heat conduction (2.16) can be simplified, for one-dimensional heat conduction without internal heat generation, to:

$$\frac{\partial^2 T}{\partial x^2} = \frac{1}{\alpha} \frac{\partial T}{\partial t} \tag{2.32}$$

This equation contains three variables and solutions are not straightforward even for simple geometries and thermal boundary conditions. Figure 2.10 shows the case of a semi-infinite solid (e.g. the surface of a large, infinitely thick plate) initially at a uniform temperature T_i that is subject to a sudden change in temperature at the surface T_0.

The heat conduction equation can be solved [4] to give the following solution for temperature T at any distance x into the solid and at any time t since the transient was applied:

$$\frac{T - T_0}{T_i - T_0} = erf \frac{x}{2\sqrt{at}} \tag{2.33}$$

where the *Error function* is given by: $erf(z) = \frac{2}{\sqrt{\pi}} \int_0^z e^{-m^2} dm$

The *Error function* varies between 0 and 1 and has the properties:

$erf(z) = 0$ when $z = 0$
$erf(z) \approx 1$ for $z > 2$

So, for all values of $\frac{x}{2\sqrt{at}}$ greater than 2, there is no change in temperature. This result then shows that the temperature profile in the solid will develop as time increases as shown in Figure 2.10. It is worth noting that (2.33) contains the *Fourier number*, described earlier as being a key dimensionless number in the analysis of transient heat conduction. The term $\frac{x}{2\sqrt{at}}$ can be written as $\frac{1}{2\sqrt{Fo}}$. So for small values of Fo, there will be no change in initial temperature, but as Fo increases so the temperature tends towards $T = T_0$. When $Fo = 1$, then $\frac{x}{2\sqrt{at}} = 0.5$ and $erf(0.5) = 0.52$, indicating that the effect of the transient will be significant at that stage.

In most practical cases, the boundary conditions are not as simple as those used in the analysis above, and usually, there will be convection at the surface of the solid. In these cases, analytical solutions become very complex and will not be practical for complex geometries. Nevertheless, the case described above illustrates the phenomena taking place and the importance of the *Fourier number* in analysing transient heat conduction. In situations where it is necessary to calculate

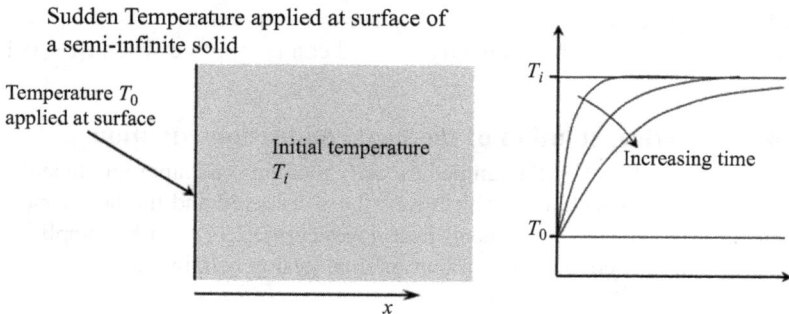

Figure 2.10   Transient heat conduction profiles in a thick solid

temperature changes during transients, it is usually appropriate to use numerical methods for solving the conduction equation.

2.2 Convection heat transfer

Convection is the transfer of heat to or from a solid surface by the movement of a fluid over the surface and the rate of heat transfer depends not only upon the thermal properties of the fluid but also on the characteristics of the fluid flow. In calculating convection heat transfer, it is essential therefore to understand the nature of the fluid flow over the heat transfer surface as well as thermal effects taking place.

The rate of heat convection from a surface (q_s) can be determined from Newton's law of cooling in terms of the surface temperature (T_s) and the fluid temperature far away from the surface (T_f):

$$q_s = hA(T_s - T_f) \tag{2.34}$$

in which A is the area of the surface and h is the *heat transfer coefficient*. It is the heat transfer coefficient that determines the rate of heat transfer by convection and its magnitude is dominated by the properties of the fluid and the nature of the fluid flow over the surface being considered.

2.2.1 Physical processes taking place in a convection

It is a fundamental principle of fluid mechanics that the molecules of a fluid in contact with a solid surface remain fixed to that surface and do not slip (the so-called *no-slip* condition). When fluid movement occurs, layers of fluid close to the surface slide over each other and the velocity increases from zero at the wall, to the mainstream velocity far away from the wall. There is therefore a layer of fluid close to the wall that is moving more slowly, known as the *hydrodynamic boundary layer*, in which there is a characteristic velocity profile. This is depicted in Figure 3.1 in Chapter 3.

In the boundary layer, as the layers of fluid slide over each other, shear forces are set up and these result in shear stress at the wall. The magnitude of the shear stress depends upon the viscosity of the fluid and the velocity gradient at the wall according to (3.3) in Chapter 3.

If a fluid flow passes over a flat surface, the boundary layer is initially thin and the shear stresses are high as the velocity gradient is steep. Further, along the surface the thickness of the boundary layer increases, the velocity gradient reduces and the shear stresses reduce.

As there is no movement of the fluid at the surface, heat can only be transferred into the fluid by conduction from the wall in the region immediately next to the wall. As heat is conducted, so the layers of fluid close to the wall increase in temperature. A *thermal boundary layer*, therefore, builds up similarly to the hydrodynamic boundary layer and thermal gradients exist within the thermal boundary layer. The rate of heat conduction can be expressed in terms of Fourier's law and the temperature gradient within the thermal boundary layer at the surface:

$$q_s'' = -k \left(\frac{\partial T}{\partial y} \right)_{y=0} \tag{2.35}$$

By combining (2.34) and (2.35), the heat transfer coefficient may be written in terms of the thermal conductivity of the fluid and the temperature gradient at the wall as follows:

$$h = -\frac{k}{(T_s - T_f)} \left(\frac{\partial T}{\partial y} \right)_{y=0} \tag{2.36}$$

So, the steeper the temperature gradient at the wall, the higher the heat transfer coefficient and the greater the rate of convection heat transfer. The thermal boundary layer may be thought of as a thermal resistance between the surface and the bulk of the fluid to which heat is being transferred. As a boundary layer develops over a surface, it becomes thicker and the thermal resistance increases. An important principle in maximizing convection heat transfer is to break up boundary layers so that they do not become too thick and impose a high thermal resistance.

2.2.1.1   Laminar and turbulent flows

At low velocities, and at the start of a boundary layer on a surface, the fluid flows in a *laminar* manner where individual particles within the fluid move in parallel paths. Under these conditions, heat transfer through the boundary layer (perpendicular to the fluid streamlines) can only take place by conduction through the layers of fluid moving over the surface. The rate of heat transfer is thus largely determined by the thermal conductivity of the fluid and the thickness of the boundary layer.

At higher fluid velocities, and as the boundary layer becomes established, the flow becomes turbulent and vortices or eddies occur within the boundary layer as the laminar flow becomes unstable. These eddies mix up the fluid in the boundary layer and increase both the transfer of heat and momentum (velocity) between the wall and the mainstream flow. The eddy motion is suppressed close to the wall and a very thin laminar boundary layer persists. However, in the bulk of the boundary layer, the mixing of the turbulent motion results in lower velocity and temperature gradients. Consequently, the velocity and temperature gradients are higher at the wall and so turbulent flow results in higher heat transfer coefficients and higher shear stresses (friction). This is illustrated in Figure 3.2 in Chapter 3, where laminar and turbulent boundary layers for flow in a pipe are shown. Note that the gradient of velocity is much higher next to the wall for turbulent flow than for laminar flow.

The laminar/turbulent nature of the flow is characterized by a non-dimensional number known as Reynolds number (Re). For a flat plate, this is given by the equation:

$$\mathrm{Re}_x = \frac{\rho u x}{\mu} \tag{2.37}$$

where x is the distance from the leading edge of the plate.

The Reynolds number is the ratio of the inertia forces to the viscous forces in a fluid flow. It is found that for Re $<$ 35,000, the flow is always laminar. There is then a transitional region where some turbulence may occur and for Re $>$ 4,000,000, the flow is fully turbulent.

For flows in pipes, the Reynolds number is conventionally based on the diameter of the pipe (d):

$$\text{Re}_d = \frac{\rho u d}{\mu} \tag{2.38}$$

It is found that laminar flow always persists for Re $<$ 2,000. However, for Re $>$ 10,000, the flow is fully turbulent once the boundary layer has become established and meet at the centre of the pipe. (*The reason for the transition to turbulence occurring at much lower values of Reynolds number for flow in pipes is due to the fact that the characteristic length for flow over a plate is the distance along with the plate, parallel to the direction of the flow, whereas for pipes it is the diameter—a dimension normal to the flow direction.*)

The features of a flow that result in high heat transfer coefficients are thus those that give steep thermal gradients in the thermal boundary layer at the wall. Thin boundary layers and high levels of turbulence both produce high heat transfer coefficients and enhance heat transfer. Thin boundary layers are promoted by features to break up the flow to prevent thicker boundary layers building up.

The temperature gradient at the wall is not an easy quantity to determine and while the fluid flow and heat conduction equations can be integrated through the boundary layer for simple flow situations, such as laminar flow along with a flat smooth plate, for more complex geometries and turbulent flows, analytical solutions for heat transfer coefficient are impractical. In engineering practice, it is normal either to use experimental correlations for convection heat transfer or to use numerical simulations for fluid flow and heat transfer. These are now readily available within CFD software.

2.2.1.2   Forced and natural convection

In considering convection heat transfer, the fluid flow may be induced in two ways. It may be driven or *forced* by an external means such as a pump or a fan. Under these conditions, the flow characteristics are normally independent of the rate of heat transfer and the heat transfer coefficient will depend predominantly upon the externally driven flow and the effect of temperature is secondary. This is known as *forced convection.*

Fluid flow may also be induced by buoyancy forces that result from the temperature differences within a fluid occurring due to heat transfer. In these situations, there is a strong coupling between the rate of heat transfer and the fluid flow, and heat transfer coefficients are found to be dependent on the temperature differences. This is known as *natural* or *free convection.*

It is important to recognize in a heat transfer situation whether the flow is driven externally or by buoyancy. In some circumstances, both forced and natural convection may occur together. This can be significant in rotating equipment where the buoyancy forces from heat transfer can be significantly enhanced due to the accelerations that occur where the fluid is rotating.

2.2.2   Correlations of heat transfer coefficient

In engineering calculations, heat transfer coefficients are determined almost exclusively from experimental correlations and the heat transfer coefficient is usually expressed in a non-dimensional form in terms of a Nusselt number. The Nusselt number is the ratio of the heat transfer by convection in a fluid to the heat transfer by conduction over a representative thickness of fluid and indicates the enhancement of convection over conduction. It is given by the equation:

$$\mathrm{Nu} = \frac{hs}{k} \tag{2.39}$$

The lowest value for Nusselt number is 1.0, which would occur in a fluid where there was no movement and the only form of heat transfer was by conduction.

2.2.2.1   Flow inside pipes and ducts

It is useful to categorize flows as either *internal* or *external*. Internal flows are those in pipes and ducts and they are characterized by the feature that as the boundary layers grow on the walls of the pipe, they eventually meet in the middle. At this point, the flow is said to be *fully developed* and there is no further change to the boundary layer profile along the pipe. The region at the start of the pipe where the boundary layers are growing is termed as the *entry region* and in this region, the heat transfer coefficient decreases as the boundary layers grow. Once the flow is fully developed, there is no further change in the heat transfer coefficient. This is illustrated in Figure 3.7 in Chapter 3.

For Reynolds numbers (based on the pipe diameter) less than about 2,000, the fully developed boundary layer will always be laminar, and the velocity profile in the pipe has a parabolic shape. For higher Reynolds numbers, the fully developed boundary layer is more likely to be turbulent and the velocity profile is much flatter.

The development of the thermal and hydrodynamic boundary layers depends on the diffusivities of heat and momentum in the fluid. Diffusivity of momentum is determined by the viscosity of the fluid—the higher the viscosity the greater the shear stresses generated at a wall and the more quickly a hydrodynamic boundary layer builds up. Diffusivity of heat is determined by the thermal conductivity – the greater the thermal conductivity the greater the heat flux into a fluid at a wall and the more quickly the thermal boundary layer builds up. The ratio between the diffusivity of momentum and the diffusivity of heat is known as the Prandtl number, defined as:

$$Pr = \frac{\mu c_p}{k} \tag{2.40}$$

Prandtl number is a key fluid property that is used extensively in convection heat transfer.

The lengths of the entry region (L) for pipe flow are given approximately by the following relationships:

For laminar flow, the hydrodynamic entry length (L_h) is:

$$L_h \approx 0.05\text{Re}\, D \qquad (2.41)$$

and the thermal entry length (L_t) is:

$$L_t \approx 0.05\text{RePr}\, D \qquad (2.42)$$

For turbulent flow, the entry length is independent of Reynolds number or Prandtl number and is given by:

$$L_h \approx L_t \approx 10D \qquad (2.43)$$

Experimental correlations for heat transfer in pipes can generally be expressed in the form:

$$Nu = f(\text{Re}, \text{Pr}) \qquad (2.44)$$

where Re is the Reynolds number for the flow in the pipe, as defined in (2.38) and Pr is the Prandtl number.

$$L_t \approx 0.05\text{RePr}\, D \qquad (2.45)$$

Some commonly used correlations for heat transfer in pipes are given below.

Laminar flow

For fully developed laminar flow ($\text{Re}_d < 2000$), it is found that the Nusselt number is constant and depends on whether the wall is at constant temperature or whether there is a constant heat flux.

For constant wall temperature in the axial direction:

$$\text{Nu}_d = 3.66 \qquad (2.46)$$

For a constant wall heat flux in the axial direction:

$$\text{Nu}_d = 4.36 \qquad (2.47)$$

For the entry region of a pipe, the following relationship may be used for constant wall temperature conditions:

$$\overline{\text{Nu}_d} = 3.66 + \frac{0.0668(d/L)\text{Re}_d\text{Pr}}{1 + 0.04[(d/L)\text{Re}_d\text{Pr}]^{2/3}} \qquad (2.48)$$

In this equation, the Nusselt number reduces as the boundary layer develops along the pipe. The expression gives the average Nusselt number from the entry of the pipe to a point a distance L into the pipe. When Nusselt numbers are stated, it is important to realize whether they are position-dependent and whether they are local or average values.

The relationship for an average Nusselt number is determined simply by integrating an equation for a local Nusselt number using the equation:

$$\overline{\text{Nu}} = \frac{1}{L}\int_0^L \text{Nu}_x dx \qquad (2.49)$$

Turbulent flow

In many cases, the fluid flow in a pipe is turbulent ($\text{Re}_d > 2,000$) and the simplest (and most widely used) expression for Nusselt number in fully developed flow is the Dittus Boelter correlation:

$$\text{Nu}_d = 0.023\text{Re}_d^{0.8}\text{Pr}^n \qquad (2.50)$$

where $n = 0.4$ when the fluid is being heated and $n = 0.3$ when the fluid is being cooled. This correlation is applicable to smooth pipes in the range $0.7 \leq \text{Pr} \leq 120$, $2,500 \leq \text{Re}_d \leq 124,000$ and $L/d > 60$. The maximum difference between experimental measurement and this correlation over its range of applicability is in the order of 40%.

If there is a significant temperature difference between the fluid and the pipe wall such that the properties of the fluid (particularly viscosity) vary significantly, then these variations can affect the behaviour of the boundary layers and hence the heat transfer coefficient. A commonly used correlation accounting for such effects is the Seider-Tate correlation given by:

$$\text{Nu}_d = 0.023\text{Re}_d^{0.8}\text{Pr}^{0.33}\left(\frac{\mu}{\mu_w}\right)^{0.14} \qquad (2.51)$$

where μ is the viscosity of the fluid at the bulk temperature and μ_w is the viscosity of the fluid at the wall temperature. This correlation is applicable to smooth pipes in the range $0.7 \leq \text{Pr} \leq 16,700$ and $\text{Re}_d > 10,000$.

When the pipe surface is rough, the heat transfer will be increased due to increased mixing in the boundary layer near the wall. A correlation that accounts for roughness effects is that of Gnielinski.

$$Nu_d = \frac{(f/2)(\text{Re}_d - 1,000)\text{Pr}}{1 + 12.7\sqrt{f/2}(\text{Pr}^{2/3} - 1)} \qquad (2.52)$$

where f is the friction factor for the duct which can be determined from a Moody chart (Figure 3.5 in Chapter 3).

Non-circular ducts

In many cases, a duct may not be circular in cross-section, and an equivalent diameter or *mean hydraulic diameter* can be used as a replacement. The mean hydraulic diameter is defined as:

$$D_h = \frac{4 \times (\text{duct cross} - \text{sectional area})}{\text{wetted perimeter of duct}} \qquad (2.53)$$

where the wetted perimeter of the duct is the perimeter of the surface in contact with the fluid.

Bulk temperature
In a pipe or duct there will be a temperature profile within fluid between the wall and the centreline. However, a bulk temperature is used (sometimes known as a mean temperature or mixing cup temperature) to determine the properties of the fluid used in the correlations for heat transfer coefficient and also in the definition of the heat transfer coefficient itself so that (2.34) becomes:

$$q_w = hA(T_w - T_b) \tag{2.54}$$

where T_w is the duct wall temperature and T_b is the bulk temperature of the fluid.

Convective heat transfer in rotating ducts
Cooling ducts are used in the rotors of some electrical machines. These ducts will usually be parallel to the axis of rotation and so there is no pressure gradient generated as a result of centrifugal forces. However, due to Coriolis effects and cross flow, there are significant effects on pressure loss in the entrance region and these are described in Section 3.5. Heat transfer relationships are given in Section 2.9.

2.2.2.2 Flow over flat plates
Laminar flow
For *constant wall temperature*, the local heat transfer coefficient over a developing laminar boundary layer is given by the equation:

$$\mathrm{Nu}_x = 0.331\mathrm{Re}_x^{0.5}\mathrm{Pr}^{0.33} \tag{2.55}$$

This may be integrated from the start of the plate to give an average heat transfer coefficient over that length given by:

$$\mathrm{Nu}_x = 0.662\mathrm{Re}_x^{0.5}\mathrm{Pr}^{0.33} \tag{2.56}$$

These equations apply for Reynolds numbers less than 500,000.

For situations with *constant heat flux* over the plate, then the following equation may be used for the local heat transfer coefficient:

$$\mathrm{Nu}_x = 0.453\mathrm{Re}_x^{0.5}\mathrm{Pr}^{0.33} \tag{2.57}$$

When integrated from the start of the plate this gives:

$$\mathrm{Nu}_x = 0.906\mathrm{Re}_x^{0.5}\mathrm{Pr}^{0.33} \tag{2.58}$$

Turbulent flow
The transition from laminar to turbulent flow occurs at a Reynolds number of about 500,000 and the following equations may be used. For $5\times10^5 < \mathrm{Re}_x < 10^7$, then the local heat transfer coefficient for constant wall temperature is given by:

$$\mathrm{Nu}_x = 0.0296\mathrm{Re}_x^{0.8}\mathrm{Pr}^{0.33} \tag{2.59}$$

This can be integrated from the start of a plate up to $5 \times 10^5 < \mathrm{Re}_x < 10^7$ to give:

$$\mathrm{Nu}_L = \left(0.037 \mathrm{Re}_L^{0.8} - 871\right) \mathrm{Pr}^{0.33} \tag{2.60}$$

In these cases, the heat transfer coefficient for a plate with constant wall flux is typically about 4% larger than the heat transfer coefficient for a plate with constant wall temperature as given here.

Film temperature

For flows over flat plates and other external surfaces, the temperature at which the fluid properties (density, viscosity, and specific heat capacity) are evaluated is the *mean film temperature* and is the mean of the wall temperature and the free stream fluid temperature away from the wall. The heat transfer coefficient is based on temperature difference between the wall and the free stream as expressed in (2.34).

Flow over external surfaces

When there is flow over an object immersed in a fluid this is generally character-ized as an external flow. The boundary layers in this situation are not confined but in many cases can become detached from the object and a region of flow recircu-lation occurs. Typical examples are objects such as cylinders, spheres, etc., as depicted in Figure 2.11.

For the case of a cylinder with a fluid flow perpendicular to the axis, the point at which the flow separates from the cylinder depends on the Reynolds number. As the heat transfer coefficient varies locally around the surface of the cylinder, the overall heat transfer coefficient varies in a complex way with the flow and an expression for heat transfer in air is given by the following for Reynolds number > 500 and fluid properties based on mean film temperature.

$$\overline{\mathrm{Nu}_d} = 0.46 (\mathrm{Re}_d)^{0.5} + 0.00128 \mathrm{Re}_d \tag{2.61}$$

For other external flow configurations, including banks of cylinders in varying array patterns, reference should be made to the extensive literature available [1–5].

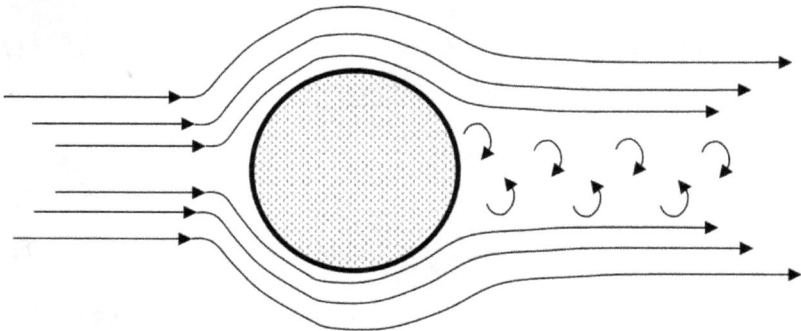

Figure 2.11   Flow over a cylinder showing separation of the boundary layer

Natural convection

Natural convection occurs when no external fan or pump is driving the fluid flow and the movement is caused by the buoyancy forces generated due to temperature differences in the fluid as a result of heat transfer. Buoyancy forces may arise due to the effects of gravity, but the accelerations that occur with rotation may also generate body forces within a fluid and induce natural convection. In natural convection, it is the temperature differences that drive the fluid motion and so there is a strong coupling between heat transfer and fluid flow. This is in contrast to forced convection where, in general, the fluid flow is independent of the heat transfer.

The boundary layer profiles have a distinct shape in natural convection as shown in Figure 2.12 with a heated wall immersed in a cooler fluid. Heat is transferred from the wall to the fluid by conduction. This causes the fluid close to the wall to expand resulting in buoyancy forces. There is no fluid motion at the wall due to the no-slip condition, and far away from the wall, there is no buoyancy as the fluid is at the bulk temperature of the surroundings. The velocity boundary layer thus shows a peak at a little distance away from the wall.

In natural convection, the fluid flow is characterized by a dimensionless number called the Grashof number (Gr). This is the ratio of the buoyancy forces to the shear forces in the flow and has the general form:

$$Gr = \frac{\beta g \rho^2 (\Delta T) l^3}{\mu^2} \qquad (2.62)$$

where β is the coefficient of cubical expansion for the fluid ($= 1/T$ for air), ΔT is the temperature difference between the surface and the bulk fluid, and l is a

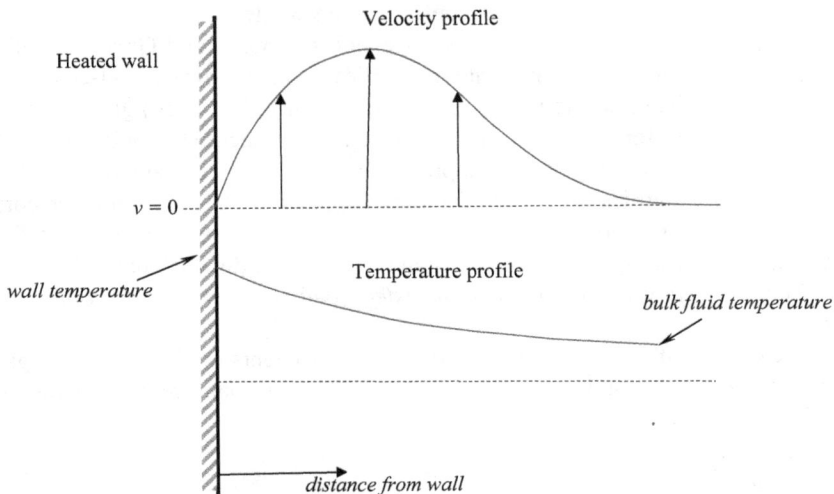

Figure 2.12   Temperature and velocity profiles in the boundary layer for natural convection from a heated vertical surface

representative dimension for the surface in the direction of the gravity vector. At low values of Grashof number, the boundary layer is laminar but a transition to turbulence occurs at $Gr \approx 10^9$.

Heat transfer coefficients for natural convection are usually determined from correlations that take the general form:

$$Nu_l = C(Gr_l Pr)^n \tag{2.63}$$

where l is some representative dimension.

The product Gr Pr is also known as the Rayleigh number (Ra).

For flat vertical surfaces, the appropriate correlations are:

$$Nu_l = 0.59(Gr_l Pr)^{0.25} \tag{2.64}$$

For $10^4 < Ra_l < 10^9$ (laminar flow):

$$Nu_l = 0.129(Gr_l Pr)^{0.33} \tag{2.65}$$

For $10^9 < Ra_l < 10^{12}$ (turbulent flow)

In each of these correlations, the Nu gives the average heat transfer coefficient over the whole vertical height of the plate l. Note that in these correlations, the Nusselt number, and hence heat transfer coefficient, is dependent (through the Grashof number) on the temperature difference between the surface and the fluid. This is a distinctive feature of natural convection, whereas, in forced convection, the heat transfer coefficient is normally independent of temperature difference.

Correlations for other geometries are given in other Refs [2–5].

2.2.2.3   Effect of variable fluid properties

Where expressions for heat transfer coefficient are written in terms of dimensionless parameters such as Nusselt number, Reynolds number, and Grashof number, then the effects of variable material properties, such as density, viscosity, are included within the dimensionless number. For instance, Eq. (2.50) giving the heat transfer coefficient for fully developed flow in a round duct, will apply equally for a liquid or a gas, as long as the appropriate fluid properties are used and the dimensionless quantities are within the range applicability for a particular correlation. However, sometimes heat transfer coefficients have been derived from measurements under specific conditions and are not non-dimensionalised.

Example – a variation of heat transfer coefficient on end windings with altitude

An example of this would be heat transfer coefficients on the end windings of electrical machines described in Section 4.1.3 where in many references the heat transfer coefficient is given by the expression:

$$h = k_1 \left[1 + k_2 (Vel)^{k_3} \right] \tag{2.66}$$

This equation has been derived from experimental work on end windings cooled by air at atmospheric pressure (at sea level). If an electrical machine were to be used in an aerospace application, then the pressure inside the end region is likely to be much lower at high altitudes and a correction would be required. To allow for the effects of altitude, an equation with a pressure correction term may be used [10], where the pressure ratio term $\left(\frac{p_z}{p_0}\right)$ is taken from the International Standard Atmosphere tables for atmospheric pressure at altitude.

$$h = k_1 \left(\frac{p_z}{P_0}\right)^{0.5} \left[1 + k_2 \left(\frac{p_z}{P_0}\right)^{(k_2 - 0.5)} \times (Vel)^{k_3}\right] \tag{2.67}$$

2.3 Radiation heat transfer

2.3.1 Nature of thermal radiation

All matter emits energy in the form of electromagnetic radiation provided that the temperature is above absolute zero. The distinguishing feature is that the transfer of radiation does not depend on the presence of an intervening medium and so radiation heat transfer is the only form of heat transfer that can take place in a vacuum.

2.3.1.1 Black body emission

The law which determines the maximum amount of radiation emitted from an object is the Stefan–Boltzman law. The maximum emission is more commonly known as the *black body* emission and is expressed as shown below:

$$q_b'' = \sigma T^4 \tag{2.68}$$

In this equation, σ is known as the *Stefan–Boltzmann* constant and has the value of 5.67×10^{-8} W/m^2 K^4. The temperature must be expressed in Kelvin and q_b'' is the rate of radiation emission in W/m^2. The Stefan–Boltzmann law is not an empirical law as the Stefan–Boltzmann constant can be determined theoretically from quantum theory. This law is characterized by the rate of radiation being proportional to temperature to the fourth power. So radiation becomes much more important as the temperature rises.

2.3.1.2 Emissivity

The *Stefan–Boltzmann* law gives the maximum amount of radiation that can be emitted from an object. It is found that the actual amount of radiation emitted by an object depends on the nature of the surface. In reality, all surfaces emit less than the black body radiation, and a property known as *emissivity* is defined to represent the proportion of the black body radiation emitted by a real surface. Emissivity varies from 0 to 1 and as a broad generalization, polished metallic surfaces have low emissivities (typically less than 0.2), whereas rough non-metallic surfaces have high emissivities (typically >0.8). The emissivity of a black body is unity by definition. The emissivity of a surface depends on the wavelength of the radiation emitted, the direction of emission, and the surface temperature. Generally, it is only polished flat surfaces that exhibit properties that vary with direction and so for many practical engineering situations, it can be assumed that surfaces behave in a *diffuse* manner and that emissivity does not vary with direction. Similarly, it is common practice to define a mean emissivity for a surface by integrating the emission over all wavelengths and expressing it as a fraction of the black body emission from a surface at the same temperature. This mean emissivity is also known as a *grey* emissivity that is assumed to be constant over all wavelengths. Values of emissivity can be found in Refs [1–5].

It is important to mention that radiation heat transfer is a phenomenon determined by the nature of the surface of an object. Any change to the surface, such as the application of a coat of paint or varnish, the presence of an oxide layer due to corrosion may significantly change the thermal radiation heat transfer effects. Uncertainty over the nature of a surface is a significant cause of uncertainty in the calculation of radiation heat transfer.

2.3.1.3   Absorption, reflection, and transmission

When radiation is incident on a surface, it may be *absorbed, transmitted, or reflected* and three properties are defined to represent these effects.

Transmissivity τ represents the proportion of radiation that is transmitted. Absorptivity α represents the proportion that is absorbed by the surface. Reflectivity ρ represents the proportion that is reflected by the surface.

For any surface, the sum of the three properties must equal unity as all the incident radiation is either transmitted, absorbed, or reflected (see Figure 2.13):

$$\tau + \alpha + \rho = 1 \tag{2.69}$$

Each of these properties is dependent upon the temperature of the surface and the direction and wavelength of the radiation. However, for practical calculations, it is usually assumed that a surface is diffuse (i.e., that the properties are independent of the direction the radiation comes from. This is true except for highly polished surfaces such as mirrors, where the angle of reflection of the radiation is equal to the angle of incidence.). For opaque surfaces, the transmissivity is zero and so (2.69) becomes:

$$\alpha + \rho = 1 \tag{2.70}$$

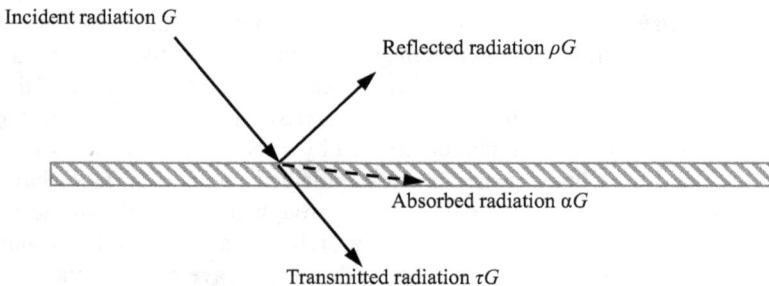

Figure 2.13   Radiation interactions with a surface

Radiation properties vary with the wavelength of the incident radiation for many substances and so the absorptivity or reflectivity of solar radiation (which is of short wavelength as the sun is very hot) may be different from the absorptivity or reflectivity of radiation from low temperature (long wavelength) surfaces. However, for most engineering applications, where the radiation is from surfaces at relatively low temperatures of say a few hundred degrees, the radiation properties can be assumed to be independent of wavelength.

2.3.1.4   Kirchoff's law of radiation

It can also be shown that for opaque objects the emissivity is equal to the absorptivity for radiation of one wavelength. This is known as Kirchoff's law and so the absorptivities are similar to the emissivities given for radiation involving low temperatures such as those in most typical electrical machines. For an opaque surface, the reflectivity can be determined from (2.70).

2.3.1.5   Liquids and gases

Dry air is transparent to thermal radiation and so does not impede the radiation heat transfer process. Some other gases, notably carbon dioxide and water vapour are not transparent but absorb radiation. The amount of absorption depends on the partial pressure of the gas and the length of the radiation path. In the air, the concentrations of water vapour and carbon dioxide are usually so low that radiation interactions are negligible. However, in the products from combustion processes, the concentrations of carbon dioxide and water vapour are much higher and gas radiation is important. In the cooling of most electrical machines, the gases involved will generally be transparent to radiation.

Liquids are generally opaque to thermal radiation and so thermal radiation need not be considered between surfaces immersed in liquids.

2.3.2   Heat transfer due to radiation

2.3.2.1   View factor

When two surfaces are in proximity, heat transfer due to radiation occurs as a result of the two-way process of the emission from one surface being absorbed by the second surface and the emission from the second surface being absorbed by the first surface. All the radiation leaving a surface passes through a hemisphere over the point of radiation and the amount of radiation that is intercepted by another surface depends on the proportion of the hemisphere that the surface obscures. A *view factor* is thus defined to represent the proportion of radiation emitted from one surface that is incident on a second surface. The view factor is dependent solely on the size and shape of the two surfaces and the geometric relationship between them and it is a quantity that varies from 0 to 1. For instance, for two surfaces that are parallel to each other and large in an area compared to the distance between them, the view factor will approach the value of 1. However, if two small surfaces are a large distance apart, then the view factor from one to the other will tend towards zero. The view factor (sometimes referred to as the radiation configuration factor or

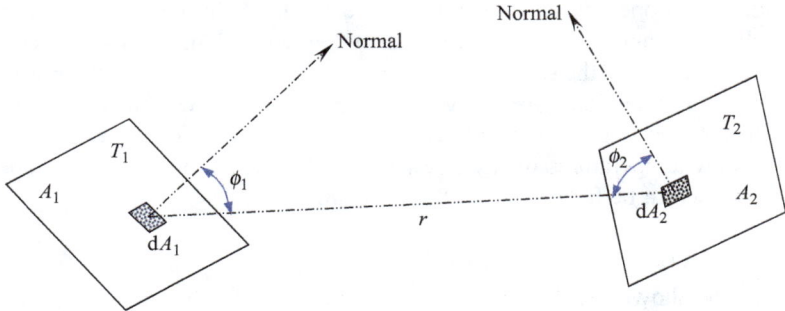

Figure 2.14   Representation of two black surfaces radiating to each other

shape factor) from one surface to another can be calculated by (2.71) below, in which the terms are as shown in Figure 2.14:

$$F_{12} = \frac{1}{A_1} \int_{A_1} \int_{A_2} \frac{\cos\theta_1 \cos\theta_2}{\pi r^2} dA_1 dA_2 \qquad (2.71)$$

In this equation, F_{12} is the view factor from surface 1 to surface 2 (i.e., the proportion of the radiation emitted from surface 1 that is incident on surface 2), A_1 and A_2 are the areas of each surface, θ_1 and θ_2 are the angles on each surface between a line between the surfaces, of length r, and the normal to each surface.

For many practical geometries, the view factors between two surfaces have been determined and may be read from graphs, and examples are given in Refs [1–5].

2.3.2.2   Relationship between view factors

The view factor is defined as the proportion of radiation from one surface that is incident on a second surface, and for radiation from the second surface back to the first surface, the view factor is not necessarily the same. However, it can be shown that there is a relationship between the view factors between two surfaces that is written as:

$$A_1 F_{12} = A_2 F_{21} \qquad (2.72)$$

This is known as a *reciprocity relationship* for view factors and can conveniently be used when calculating heat transfer between two surfaces. It shows that the view factors between two surfaces F_{12} and F_{21} are only equal when the areas of the two surfaces are the same.

It can also be shown that for many surfaces making an enclosure, the sum of the view factors from one surface to each of the others must be equal to 1. Thus,

$$\sum_{j=1}^{N} F_{i \to j} = 1 \qquad (2.73)$$

So, for several surfaces in an enclosure, if some of the view factors can be determined from graphs, unknown view factors may be deduced by using this summation rule.

A further summation rule can also be derived in terms of the view factor between one surface and many other smaller surfaces which make up a large surface:

$$F_{12} = \sum_{j=1}^{N} F_{1 \rightarrow j} \tag{2.74}$$

where surfaces $j = 1$ to N are small sections that make up the larger surface 2.

2.3.2.3   Heat transfer between black surfaces

Radiation heat transfer between two surfaces occurs when the radiation heat flows in each direction between the two surfaces are out of balance and a net heat flow occurs. The radiation heat flow from one black surface to another is equal to the total black body emission from the surface multiplied by the view factor with respect to the other surface:

$$q_{1 \rightarrow 2} = A_1 F_{1-2} \sigma T_1^{\,4} \tag{2.75}$$

The radiation that the first surface receives from the second surface is correspondingly given by:

$$q_{2 \rightarrow 1} = A_2 F_{2-1} \sigma T_2^{\,4} \tag{2.76}$$

The heat transfer is the difference between these two heat flows and by using the *reciprocity relationship* (2.72), the heat transfer can be expressed as:

$$q_{12} = q_{1 \rightarrow 2} - q_{2 \rightarrow 1} = A_1 F_{1-2} \sigma \left(T_1^{\,4} - T_2^{\,4} \right) \tag{2.77}$$

2.3.2.4   Heat transfer between grey surfaces

When the surfaces involved in heat transfer are not black but have emissivities less than unity, the situation becomes more complex as incident radiation is reflected from surfaces as well as being absorbed. So, the radiation leaving a surface is not just the emitted radiation but also includes reflected incident radiation, as illustrated in Figure 2.15. A new term is defined called *radiosity* (J) which represents the total radiation leaving a surface, as expressed in the following equation:

$$J = \varepsilon \sigma T^4 + \rho G \tag{2.78}$$

The net radiation heat transfer from a surface q is thus the difference between the incident radiation (G) and the radiosity (J).

By expressing the radiosity from each surface in terms of the temperature, the heat transfer between two grey surfaces q_{12} can be expressed by (2.79). The full analysis for the derivation of this equation is given in Ref. [1].

$$q_{12} = \frac{\sigma \left(T_1^{\,4} - T_2^{\,4} \right)}{\left(\frac{1-\varepsilon_1}{A_1 \varepsilon_1} + \frac{1}{A F_{1-2}} + \frac{1-\varepsilon_2}{A_2 \varepsilon_2} \right)} \tag{2.79}$$

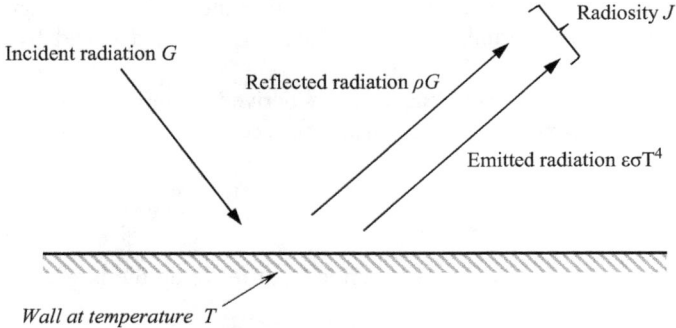

Figure 2.15   Representation of radiosity from a surface

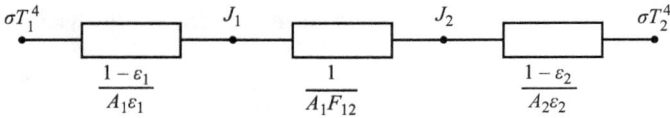

Figure 2.16   Thermal resistances between two surfaces radiating to each other

This equation may be considered in terms of thermal resistances as shown in Figure 2.16 where the terms containing the emissivities represent thermal resistances at the two surfaces and the term containing the view factor represents the thermal resistance due to the geometrical relationship between the surfaces.

This thermal resistance approach can be extended to describe radiation heat transfer between any number of grey surfaces in which the surface resistance is expressed as $\frac{1-\varepsilon}{A\varepsilon}$ and the view factor resistance between any two surfaces is expressed as $\frac{1}{AF_{1-2}}$.

There are several special cases where (2.79) can be simplified.

2.3.2.5   Large parallel surfaces

If two surfaces are large in relation to the distance between them and parallel to each other, then $A_1 \approx A_2$ and $F_{1\text{-}2} = 1$. In an electrical machine, this could be the case for radiation heat transfer across the air gap. Equation (2.79) simplifies to:

$$q_{12} = \frac{A_1 \sigma \left(T_1^{\,4} - T_2^{\,4}\right)}{\left(\frac{1}{\varepsilon_1} + \frac{1}{\varepsilon_2} - 1\right)} \tag{2.80}$$

2.3.2.6   Small surface in a large enclosure

If a small surface (1) is surrounded by a large enclosure (2), then $F_{1\text{-}2} = 1$ and $A_1 << A_2$ and so $A_1/A_2 \approx 0$. Equation (2.79) then simplifies to:

$$q_{12} = \varepsilon_1 A_1 \sigma \left(T_1^{\,4} - T_2^{\,4}\right) \tag{2.81}$$

2.3.3 Combined radiation and convection (the radiation heat transfer coefficient)

In many problems involving convection heat transfer and air, radiation heat transfer also takes place. As thermal radiation is a non-linear phenomenon and varies with absolute temperature to the fourth power, it is not easy to solve problems involving combined convection and radiation. However, the radiation heat transfer equation can be rearranged to derive an effective radiation heat transfer coefficient that can be used to simplify the analysis.

Heat transfer by convection from a wall to a fluid q_{conv} is given by (2.34) as:

$$q_{conv} = h_s A_s (T_s - T_\infty) \tag{2.82}$$

Radiation heat transfer q_{rad} from a wall to a surrounding black enclosure (where the view factor $F_{1-2} = 1$) at the same temperature as the fluid is given by:

$$q_{rad} = \varepsilon_s h_s A_s \left(T_s^4 - T_\infty^4\right) \tag{2.83}$$

This may be re-written in terms of a radiation heat transfer coefficient (also refer to section 4.1) as:

$$q_{rad} = h_{rad} A_s (T_s - T_\infty) \tag{2.84}$$

where

$$h_{rad} = \varepsilon_s \sigma (T_s + T_\infty)\left(T_s^2 + T_\infty^2\right) \tag{2.85}$$

In the case where the view factor (F_{1-2}) is less than one and the enclosure cannot be assumed to be *black*, then, incorporating other terms from (2.79) gives:

$$h_{rad} = \frac{\sigma(T_s + T_\infty)\left(T_s^2 + T_\infty^2\right)}{\left(\left(\frac{1-\varepsilon_s}{\varepsilon_s}\right) + \frac{1}{F_{S-enc}} + \frac{A_s}{A_{enc}}\left(\frac{1-\varepsilon_{enc}}{\varepsilon_{enc}}\right)\right)} \tag{2.86}$$

where the subscript $_{enc}$ refers to the enclosure and assuming that the enclosure is at the same temperature as the fluid ($T_{enc} \approx T_\infty$).

By expressing the radiation heat transfer in terms of a heat transfer coefficient (h_{rad}), it becomes much easier to calculate the combined effects of radiation and convection and the total heat flow from a surface can then be determined by combining the convection and radiation thermal resistances to give:

$$q_{total} = q_{conv} + q_{rad} = (h_{rad} + h_{conv})A_s(T_s - T_\infty) \tag{2.87}$$

Clearly, the radiation heat transfer coefficient is strongly dependent on temperature and if the temperature differences are large, then a calculation scheme may need to be iterative to account for the variation in h_{rad} with temperature.

2.3.3.1 Importance of radiation heat transfer

Where radiation and convection heat transfer from a surface take place together, it is worth considering the magnitude of each heat flow to establish whether one

Table 2.3   Typical values of radiation heat transfer
coefficient (h_{rad})

Temperature (°C)	h_{rad} (W/m² K)
0	4.6
25	6.0
50	7.6
100	11.8
150	17.2
250	32.4

mode of heat transfer is dominant. This can be done by comparing the magnitude of the heat transfer coefficients. This has been done in Table 2.3.

The radiation heat transfer coefficient has been evaluated from (2.86) assuming a black surface ($\varepsilon = 1$) and assuming that the temperature difference is small so that $T_s \approx T_\infty = T$.

It is useful to compare these values of radiation heat transfer coefficient with typical convection coefficients for air. For natural convection, typical heat transfer coefficients are in the range of 3–15 W/m² K for temperature differences up to 100 °C and surfaces up to 1 m in height. For forced convection heat transfer in ducts, typical values are in the range 10–200 W/m² K for air velocities in the range 5–50 m/s in smooth ducts of 10 mm to 1 m diameter. From this, it can be seen that in situations involving natural convection, heat transfer by radiation will also usually be significant.

However, if there is forced convection, the heat transfer coefficients will typically be an order of magnitude larger and will dominate radiation heat transfer effects. In ventilated electrical machines, it is customary to ignore the effects of thermal radiation and a design based on convection heat transfer will thus be suitably conservative.

2.4   Extended surfaces (fins)

2.4.1   Introduction

In designing heat dissipation systems, it is often found that the rate of convection heat transfer from a surface to the air limits the heat dissipation rate that can be achieved without incurring excessively high surface temperatures. In such circumstances, the rate of convection heat transfer can be increased by extending the area of the surface using fins. Examples of some extended surface geometries are given in Figure 2.17.

In designing fins, there are several important considerations. First, the geometry of the fins should not significantly reduce the convection heat transfer coefficient. This means that the fins should generally be aligned with the airflow direction and that the fins should not be spaced too close together so that the

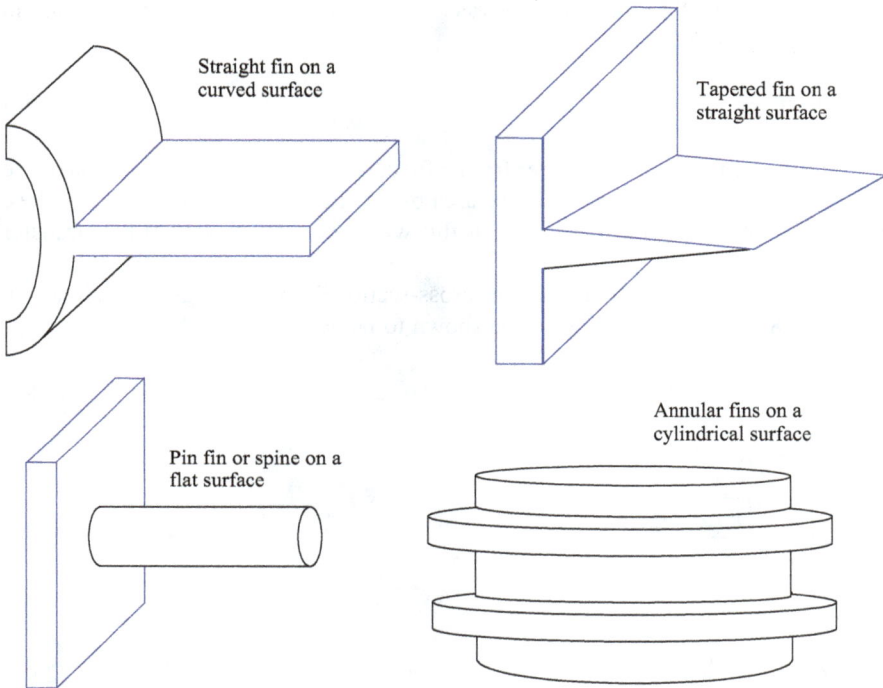

Figure 2.17   Typical fin geometries used to increase heat transfer by convection

boundary layers on adjacent fins interfere with each other. As the heat convected from the surface of the fin has to be conducted along the fin from its root, it is equally important that the conduction thermal resistance is not so great that there are unacceptable temperature gradients along the length of the fin. A lower surface temperature towards the end of the fin will reduce the rate of heat convection as the temperature difference between the fin surface and the air will be lower.

The performance of a fin can be analyzed by considering the heat conduction along the fin combined with the heat convection from its surface. And analytical solutions for fins can be found in most heat transfer textbooks: Simonson [1], Holman [2], Bergman *et al.* [3], and Bejan [4]. These solutions are straightforward if the fins are *straight,* that is to say, they have a constant cross-sectional area along their length, but when the fin is tapered or when radial fins are placed on cylinders, then the solution is more complex. However, for practical purposes, the perfor-mance of a fin can be defined in terms of *fin efficiency.*

2.4.2   Fin efficiency

Fin efficiency is a dimensionless parameter and is the ratio of the actual heat transfer from a fin (q_{fin}) to the heat transfer from the fin if it were all at the temperature of the fin root and there was no temperature gradient along the fin. Fin efficiency (η_f) is defined as:

$$n_f = \frac{q_{fin}}{hA_f(T_{root} - T_\infty)} \tag{2.88}$$

The actual rate of heat transfer from a fin can thus be calculated in terms of the fin efficiency and the exposed surface area of the fin (A_f). A fin efficiency as close to 100% as possible is most desirable as this will maximize the rate of heat transfer from the fin.

For a rectangular fin of constant cross-sectional area (straight), as shown in Figure 2.18, the fin efficiency can be shown to be approximated by:

$$n_f = \frac{\tanh(mL_c)}{mL_c} \tag{2.89}$$

where m is given by:

$$m = \sqrt{\frac{hp}{kA_c}} \tag{2.90}$$

and

$$L_c = L + \frac{A_c}{p} \tag{2.91}$$

For a wide fin ($w \gg t$) then $L_c \sim L + t/2$ and $= \sqrt{\frac{2h}{kt}}$.

Rectangular fins are most commonly used, for example, on the outside of a fan-cooled totally enclosed machine, and fin efficiency for a wide straight rectangular fin may be expressed graphically as shown in Figure 2.19, in which $m = \sqrt{\frac{2h}{kt}}$.

For many other common fin geometries, fin efficiency has been calculated and is presented in graphical form in Refs [2–5].

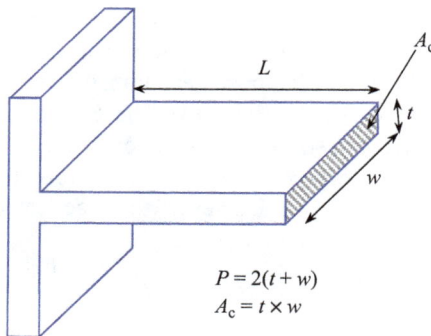

$$P = 2(t + w)$$
$$A_c = t \times w$$

Figure 2.18   A straight rectangular fin

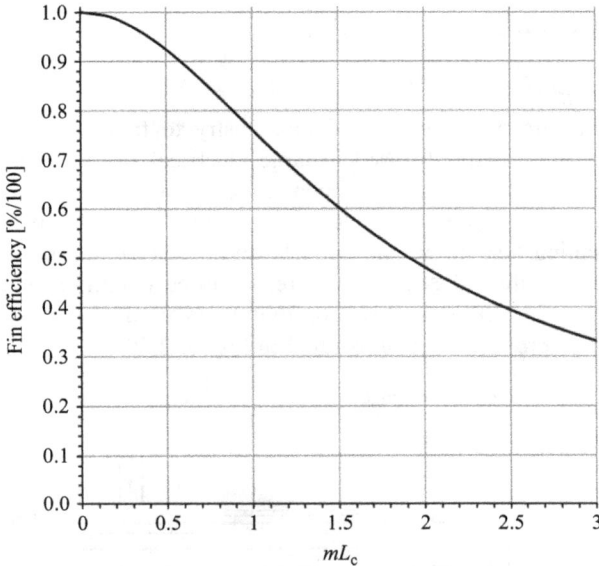

Figure 2.19 Efficiency for a wide straight rectangular fin

This analysis is only applicable for plate fins of constant rectangular cross-section in which the heat transfer along the fin is predominantly one-dimensional. That is to say, the temperature gradient through the thickness of the fin is small compared to the temperature difference between the fin surface and the surrounding air. This is true provided that the Biot number based on fin thickness is small, i.e.:

$$\sqrt{\frac{ht}{k}} \ll 1 \tag{2.92}$$

2.4.3 Fin effectiveness

An alternative performance indicator for fins is *fin effectiveness* (ε_f). This is the ratio of the actual heat transfer from a fin to the heat transfer from the root area of the fin if the fin was not there. Clearly, it is desirable for the fin effectiveness to be much greater than 1 so that the heat transfer from a surface is significantly enhanced by adding fins. The fin effectiveness is given by:

$$\varepsilon_f = \frac{q_{fin}}{hA_f(T_{root} - T_\infty)} = n_f \frac{A_f}{A_c} \tag{2.93}$$

The total rate of heat transfer from a surface with fins is then evaluated by considering the performance of each fin and the number of fins placed on the surface, not forgetting that there is also heat transfer from the surface in between the fins.

2.5   Heat exchangers

2.5.1   Introduction

Heat exchangers are extensively used in industry to transfer heat between two fluids and they are commonly used in large electrical machines to transfer heat from an internal coolant circuit to a secondary coolant such as air or water. These heat exchangers are normally of the *recuperator* type, where the two fluid streams that are exchanging heat flow continuously and are separated by a wall through which the heat transfer takes place. There are three common recuperator types corresponding to the different flow configurations used; these are *counterflow*, *parallel flow*, and *cross-flow,* as illustrated in Figure 2.20.

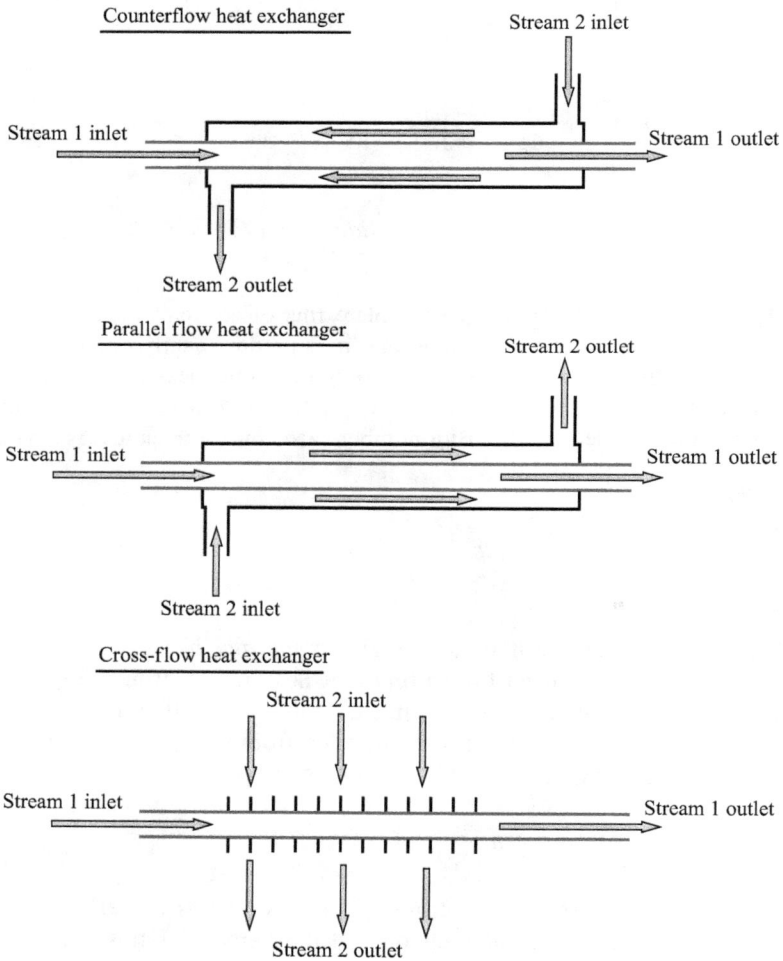

Figure 2.20   Heat exchanger configurations

The most fundamental analysis of a heat exchanger is an energy balance in which the rate of heat transfer between the two streams (q) is equated to the enthalpy lost by the hot stream and enthalpy gained by the cold stream, assuming that there are no other heat losses. The energy balance is thus:

$$\left(\dot{m}c_p\right)_{hot}(T_{hi} - T_{ho}) = \left(\dot{m}c_p\right)_{cold}(T_{ci} - T_{co}) = q \qquad (2.94)$$

Usually, the inlet temperatures of the two streams are known and so another equation is needed to be able to calculate the two outlet temperatures.

The temperature changes in the hot and cold streams are shown for a counterflow and parallel flow heat exchanger in Figure 2.21. As the size of a heat exchanger increases, there is more heat transfer and the temperature difference between the streams at the outlet is smaller. But the lines can never cross as this would imply that heat was being transferred from cold to hot.

2.5.2   Overall heat transfer coefficient

The rate of heat transfer between the two streams can be determined from the temperature difference between the streams (ΔT_m), the area of the heat exchanger surface separating the streams (A), and an overall heat transfer coefficient (U) between the two streams. It is given by:

$$q = UA(\Delta T_m) \qquad (2.95)$$

The temperature difference between the two streams varies throughout the heat exchanger and so a mean temperature difference must be used. The derivation of an appropriate mean temperature difference (ΔT_m) is given in most heat transfer textbooks of Simonson [1], Holman [2], Bergman *et al.* [3], Bejan [4], and Bejan and Kraus [5], and can be shown to be a *logarithmic mean temperature difference* (ΔT_{lm}) defined as:

$$\Delta T_{lm} = \frac{\Delta T_i - \Delta T_o}{\ln(\Delta T_i/\Delta T_o)} \qquad (2.96)$$

where ΔT_i is the temperature difference between the two streams at the hot inlet and ΔT_o is the temperature difference between the two streams at the hot outlet. The equation applies to both counterflow and parallel flow heat exchangers, as shown in Figure 2.21. Cross flow heat exchangers can also be analyzed using (2.96) multiplied by a correction factor. Further details are given in Refs [1–5].

The overall heat transfer coefficient is determined by considering the thermal resistances to heat flow between the two streams through the wall of the heat exchanger. If the heat exchanger wall is flat, then the overall heat transfer coefficient is given by:

$$U = \left(\frac{1}{h_h} + \frac{\Delta x}{k} + \frac{1}{h_c}\right)^{-1} \qquad (2.97)$$

where h_h and h_c are the convective heat transfer coefficients between the wall and the hot stream and cold stream, respectively, Δx is the thickness of the wall, and k is the thermal conductivity of the wall.

Counterflow heat exchanger

Parallel flow heat exchanger

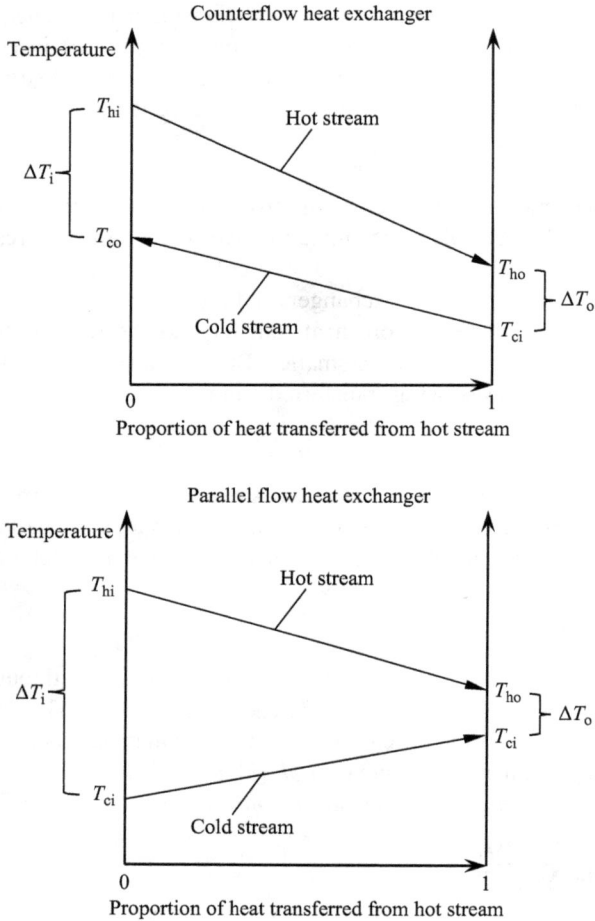

Figure 2.21   Temperature changes in the hot and cold streams as they pass through a heat exchanger. (Note that temperature lines are straight in this figure illustrating a case where the specific heat capacity of each fluid is constant. In general, the lines would be curves if specific heat capacity varied with temperature.)

If the heat exchanger wall is a tube, then the area A in (2.95) must be based either on the inside or outside area of the tube wall and the overall heat transfer coefficient must be based on whichever area is used. Equation (2.97) then becomes, when based on the internal surface area of the tubes:

$$U_i A_i = \left(\frac{1}{A_i h_i} + \frac{log_e \left(\frac{r_o}{r_i} \right)}{2\pi kL} + \frac{1}{A_o h_o} \right)^{-1}$$

(2.98)

where h_i and h_o are the heat transfer coefficients in the inside and outside of the tube walls.

In the case where the heat exchanger wall is thin and made of a high thermal conductivity material, then both (2.97) and (2.98) simplify to:

$$U = \left(\frac{1}{h_h} + \frac{1}{h_c} \right)^{-1}$$ (2.99)

The performance of a heat exchanger can thus be evaluated using (2.94) and (2.95). However, as (2.95) contains the logarithmic mean temperature difference, the equations cannot be solved explicitly but an iterative approach is usually required. This can be avoided and the analysis simplified by using an alternative approach known as the *effectiveness-number of transfer unit (NTU)* method.

2.5.3 *Effectiveness-NTU approach*

Two new quantities need to be defined for use in the effectiveness-NTU method: *capacity ratio* and *effectiveness*.

2.5.3.1 Capacity ratio *C*

The thermal capacity of a stream is the product of the mass flow rate and specific heat capacity and the capacity ratio for a heat exchanger is the ratio between the capacities of each stream. The capacity represents the amount of heat needed to increase the temperature of a stream by 1 °C.

Capacity ratio is always defined as being less than 1, so it is expressed with the larger capacity as the denominator.

$$\text{If } (\dot{m}c_p)_h > (\dot{m}c_p)_c, \text{ then } C = \frac{(\dot{m}c_p)_c}{(\dot{m}c_p)_h}$$

$$\text{If } (\dot{m}c_p)_c > (\dot{m}c_p)_h, \text{ then } C = \frac{(\dot{m}c_p)_h}{(\dot{m}c_p)_c}$$

The capacity ratio is important in determining the maximum amount of heat recovery in a heat exchanger.

2.5.3.2 Effectiveness *ε*

The effectiveness (ε) of a heat exchanger is the ratio:

$$\varepsilon = \frac{\text{Actual heat transferred}}{\text{Maximum possible heat transfer}}$$ (2.100)

In a counterflow heat exchanger, the outlet temperature of the stream with the lower capacity will approach the inlet temperature of the other stream as the heat transfer between the two streams increases. Thus, the maximum possible heat transfer is the capacity of the stream with the lower capacity multiplied by the temperature difference between the two inlet streams. The maximum heat transfer is thus $(\dot{m}c_p)_c(T_{hi} - T_{ci})$ for a case where $(\dot{m}c_p)_h > (\dot{m}c_p)_c$ (as $T_{co} = T_{hi}$). The actual heat transfer is $(\dot{m}c_p)_c(T_{co} - T_{ci})$, so the effectiveness is:

$$\varepsilon = \frac{(T_{ci} - T_{co})}{(T_{hi} - T_{ci})} \text{ if } (\dot{m}c_p)_h > (\dot{m}c_p)_c \tag{2.101}$$

If $(\dot{m}c_p)_c > (\dot{m}c_p)_h$, then the maximum possible heat transfer would occur when $T_{ho} = T_{ci}$ and will be equal to $(\dot{m}c_p)_h(T_{hi} - T_{ci})$. The actual heat transfer is $(\dot{m}c_p)_h(T_{hi} - T_{ho})$ and so

$$\varepsilon = \frac{(T_{hi} - T_{ho})}{(T_{hi} - T_{ci})} \text{ if } (\dot{m}c_p)_c > (\dot{m}c_p)_h \tag{2.102}$$

An alternative way of expressing effectiveness is:

$$\varepsilon = \frac{\Delta T(\text{fluid with minimum capacity})}{\text{Maximum temperature difference in heat exchanger}} \tag{2.103}$$

2.5.3.3   Heat transfer units

By equating (2.94) and (2.95), it is possible to express the effectiveness of a heat exchanger simply in terms of the capacities, overall heat transfer coefficient, and area. This is useful as the expression for effectiveness only contains one unknown temperature and makes analysis of a heat exchanger much easier.

Using the substitution:

$$\frac{UA}{(\dot{m}c_p)_{min}} = \text{ number of transfer units (NTU)} \tag{2.104}$$

Then for a *parallel-flow heat exchanger*, it can be shown that:

$$\varepsilon = \frac{1 - e^{-NTU(1-C)}}{1 + C} \tag{2.105}$$

where C is the capacity ratio.

A similar analysis for a *counter-flow heat exchanger* gives:

$$\varepsilon = \frac{1 - e^{-NTU(1-C)}}{1 - Ce^{-NTU(1-C)}} \tag{2.106}$$

Expressions can also be derived for other heat exchanger configurations [1,2,3,4,5].

2.6   Convective heat transfer enhancement

Convective heat transfer at a solid surface can be increased if the boundary layer is thinner and there are steeper temperature gradients between the surface and the

Transverse rib roughness elements

Wire coil insert roughness element

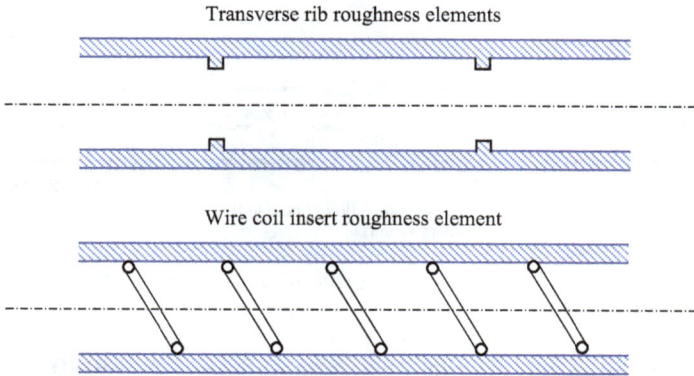

Figure 2.22   Some roughness elements that may be used to enhance heat transfer

bulk fluid. This can be achieved by using devices to encourage disruption and breakup of the boundary layer. Passive devices include the enhancement of the roughness of a surface or in the case of flow in pipes, using devices to produce a swirl flow. Some examples of features that may be used to passively enhance heat transfer are shown in Figure 2.22. The disadvantage of these devices is that they increase the pressure loss in the flow and this usually results in a reduction in the fluid flow or they may require more power input to a fan or cooling liquid pump.

The selection and design of heat transfer enhancement devices are complex and further information is provided in Ref. [5]. Their performance can be modelled using CFD.

Active systems can also be used to enhance convective heat transfer by, for instance, inducing vibrations in the surface or the fluid. Jet impingement is a technique often used and an example of this is described in Chapter 7, where impingement of the cooling liquid is used to increase the heat transfer from the end windings in a flooded stator.

2.7   Heat transfer with a phase change (heat pipes)

When convection heat transfer takes place with a change in phase, very much larger heat transfer coefficients can be achieved than are possible with single-phase flow. This happens because very large enthalpy changes take place under isothermal conditions when liquids boil or condense and so large quantities of heat can be transferred if boiling or condensation takes place. This phenomenon has been utilized in devices known as heat pipes. These are evacuated hollow tubes that contain a small quantity of a volatile liquid as illustrated in Figure 2.23. When heat is applied to one end of the heat pipe, the liquid evaporates and the vapour moves to the cooler end where it condenses. The liquid then travels either by gravity to the hotter end, if the hot end is below the cool end, or by capillary action against gravity along with a wick inside the heat pipe. As the liquid returns from the cold

Heat input

Heat output

Vapour flow

Liquid flow

Hot end (evaporator)

Cold end (condenser)

Wick structure (allows liquid to flow by
capillary action)

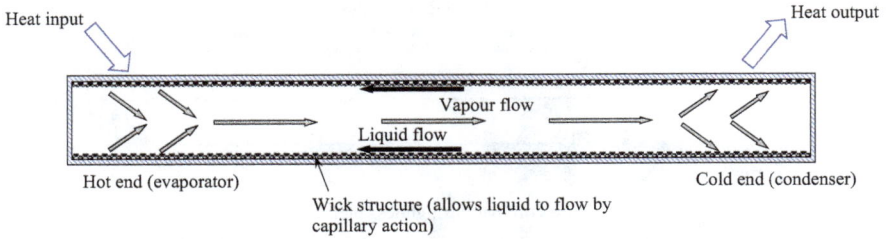

Figure 2.23   Diagram of a heat pipe

end to the hot end of the heat pipe, heat pipes work more effectively if the liquid movement is assisted by gravity. So, they work best when the hot end is lower than the cold end or where there is an acceleration from rotation and the hot surface is at a greater radius than the cold surface.

As the vapour travels along the heat pipe without any significant temperature gradient, the effective thermal conductivity of a heat pipe can be several orders of magnitude greater than the best solid materials (e.g. copper). Often the greatest thermal resistance in the use of a heat pipe is the resistance encountered in getting the heat into and out of the heat pipe at each end. An extensive coverage of heat pipes is given in Ref. [5].

Heat pipes are widely used in the cooling of electronics and most laptop computers use a heat pipe to transfer heat from the central processor chip to a finned heat sink cooled with air blown by a fan. They have been considered for use in electric machines, for example, by placing one along the centre of the shaft to conduct heat more effectively from the rotor. If the motor is fan-cooled, the fan can act as an effective heat sink to dissipate the heat to the ambient air.

2.8   Heat transfer in an annular gap

Convective heat transfer across the air gap in electrical machines is usually significant and in radial flux machines, the air gap is an annulus. Recent reviews of literature in this area are presented in Refs [11,12] in which much of the earlier literature on the topic is referred to. There is convection driven by the rotation of the rotor and there may also be an imposed axial flow if there is ventilation between the end regions. In an annular gap with a rotating inner cylinder, the flow regimes are described in Section 3.4. Above a critical rotational speed, toroidal vortices are formed, known as Taylor vortices, and these enhance the convective heat transfer. The transition to Taylor vortices is characterized by the Taylor number (Ta) which is defined as:

$$\text{Ta} = \frac{\rho \omega_m r_m^{0.5} s^{1.5}}{\mu} \tag{2.107}$$

where s is the air gap width ($r_o - r_i$), r_m is the mean rotor–stator radius, i.e., $r_m = (r_i + r_o)/2$, r_i and r_o are inner and outer radii of the air gap, ω_m is the rotational speed, μ is the dynamic viscosity of the fluid, and ρ is the fluid density. The critical Ta number at which the Taylor vortices are formed is approximately 41. It should be mentioned that if the outer cylinder rotates, then Taylor vortices are not formed—the case with an inner stator and outer rotor. The addition of an axial flow within the annular gap increases the rate of convection heat transfer.

Convective heat transfer in an annular gap can be described in terms of the heat transfer across the air gap from the rotor to the stator, and this is the general convention when there is no axial flow. However, when there is an axial flow, it is usual to express the heat transfer from the inner or outer surfaces to the air within the annulus.

Radiation heat transfer across the air gap may also be significant and this can be calculated using (2.80), defined earlier.

The following definitions are used:

$$\text{Nu} = h\frac{2s}{k} \tag{2.108}$$

The definition of heat transfer coefficient depends on whether heat transfer across the surfaces in the air gap (h_o) is being considered or the heat transfer from one surface to the fluid in the air gap (h_f) is considered:

$$h_o = \frac{Q''}{T_r - T_s} \tag{2.109}$$

$$h_f = \frac{Q''}{T_f - T_w} \tag{2.110}$$

where subscript r refers to the rotor, s refers to the stator, f refers to the fluid within the annular gap, and w refers to one of the walls in the annular gap.

2.8.1 *Annular gap with no axial flow*

For an annular gap with a Taylor number above but close to critical, then a modest axial flow (Reynolds number in the laminar region) can suppress the Taylor vortices and reduce heat transfer. Higher axial flow rates and Taylor numbers then cause the heat transfer to increase. At high axial Reynolds numbers in the turbulent region above 10^4, the following correlations taken from Kuzay and Scott [13] may be used for heat transfer from the rotating surface to the fluid flowing within the annulus:

For an annulus with no rotation, the Nusselt number is given by:

$$\text{Nu}_f = 0.022\text{Re}^{0.8}\text{Pr}^{0.5} \text{ for } 104 < \text{Re} < 105 \tag{2.111}$$

Defining a "rotation parameter" (ζ) as:

$$\zeta = 2b\omega/\pi V_a \tag{2.112}$$

The heat transfer in an annulus with rotation and axial flow (Nu_{fr}) can be defined as:

$$\text{Nu}_{fr} = \text{Nu}_f\left(1 + \zeta^2\right)^{0.8714} \text{ for } 104 < \text{Re} < 105 \tag{2.113}$$

For other conditions, a range of correlations are presented in Ref. [12].

2.8.2   *Effect of slotted surfaces in the annular gap*

In electrical machines, the gaps between the teeth within the stator generally result in the stator having a slotted surface rather than a smooth one. The rotor meanwhile for many machine types will have a smooth outer surface. There have been many studies on the effects of slots on heat transfer but it is difficult to draw overall conclusions due to a large number of possible geometric configurations [12]. Some studies show a slight decrease in heat transfer when there is laminar flow in the air gap, but in general, slots in the stator or rotor increase the heat transfer, particularly when the flow is turbulent and increase in convective heat transfer coefficient of up to 50% have been reported [12]. For many situations, the most appropriate approach, given the advances in the use of CFD, would be to undertake CFD modelling to understand the effects of slots in the air gap on flow and heat transfer.

2.9   Heat transfer in rotating ducts

Ducts may be placed in the rotor of electrical machines to provide cooling. These ducts will usually be parallel to the axis of rotation of the machine and the centrifugal forces due to rotation, combined with the differences in temperature generate secondary flows, perpendicular to the flow direction in the duct, that can increase heat transfer. Flow in rotating ducts is described in Section 3.5 in Chapter 3.

Due to density differences caused by heat transfer within the ducts, there may be secondary flows produced. These are more significant in laminar flows, but in turbulent flow, the usual case for air flowing in ducts in electrical machines, the effects are less important and heat transfer relationships for stationary ducts can be used for basic, conservative, calculations.

For more detailed relationships for turbulent, fully developed flow in circular ducts, the following equations [14] may be used:

$$\frac{\mathrm{Nu}_b}{\mathrm{Nu}_o} = 1.367 \left[\frac{\mathrm{Pr}^{0.67}}{\mathrm{Pr}^{0.67} - 0.050} \right] \left[1 + \frac{0.0286}{X^{0.2}} \right] X^{0.05} \tag{2.114}$$

where Nu_b is the Nusselt number for the rotating duct based on bulk fluid properties.

$\mathrm{Nu}_o = 0.038 \mathrm{Re}^{0.75} \mathrm{Pr}^{0.33}$ (Nusselt number for the stationary duct).

$$X = \frac{\mathrm{Ra}_\tau}{\mathrm{Re}^{2.273} \mathrm{Pr}^{0.606}} \tag{2.115}$$

$$\mathrm{Ra}_\tau = \frac{\Omega^2 H \beta T a^3}{a \upsilon} \text{ (rotational Rayleigh number)} \tag{2.116}$$

- Ω is the angular velocity of the rotor; H is the radial distance of the duct from the axis of rotation; a is the duct diameter; β is the coefficient of cubic expansion for the fluid; ΔT is the temperature difference between the duct wall and fluid; α is the fluid thermal diffusivity; v is the fluid kinematic viscosity; Re is the Reynolds number for the fluid axial flow in the duct; Pr is the fluid Prandtl number.

For short ducts where entrance effects are significant, the Coriolis force causes swirl at the entry and this increases the heat transfer coefficient in the entry region. Further information can be found in Ref. [14].

References

[1]   Simonson J. R. *Engineering Heat Transfer*, 2nd edn. UK: Palgrave MacMillan; 1988.

[2]   Holman J. P. *Heat Transfer*, 10th edn., New York, NY: McGraw-Hill; 2019.

[3]   Bergman T. L., Lavine A. S., Incropera F. P., and Dewitt D. P. *Fundamentals of Heat Transfer*, 7th edn., New York, NY: John Wiley and Sons; 2011.

[4]   Bejan A. *Heat Transfer*, New York, NY: John Wiley and Sons; 1993.

[5]   Bejan A. and Kraus A. D. *Heat Transfer Handbook*, Hoboken, NJ: John Wiley and Sons; 2003.

[6]   Motor-CAD Reference Manual.

[7]   Bolton W. *Newnes Engineering Materials Pocket Book*, 2nd edn., New York, NY: Elsevier; 1996.

[8]   Staton D., Boglietti A., and Cavagnino A. 'Solving the more difficult aspects of electric motor thermal analysis'. *IEEE International Electric Machines and Drives Conference*, Madison, WI, USA, June 2003, pp. 747–755.

[9]   Kulkarni D. P., Rupertus G., and Chen E. 'Experimental investigation of contact resistance for water cooled jacket for electric motors and generators'. *IEEE Trans. Energy Convers.* 2012; 27(1):204–210.

[10] Gilson G., Pickering S. J., Hann D. B., and Gerada C. 'Analysis of the end winding heat transfer variation with altitude in electric motors'. *35th Annual Conference of IEEE Industrial Electronics*, Porto, Portugal, November 2009, pp. 2545–2550.

[11] Howey D. A., Childs P. R. N., and Holmes A. S. 'Air-gap convection in rotating electrical machines'. *IEEE Trans. Ind. Electron.* 2012; 59(3):1367–1375.

[12] Fénot M., Bertin Y., Dorignac E., and Lalizel G. 'A review of heat transfer between concentric rotating cylinders with or without axial flow'. *Int. J. Therm. Sci.* 2011;50(7):1138–1155.

[13] Kuzay T. M. and Scott C. J. 'Turbulent heat transfer studies in an annulus with inner cylinder rotation'. *Trans. ASME J. Heat Transfer* 1977; 99(2):12–19.

[14] Morris W. D. *Heat Transfer and Fluid Flow in Rotating Coolant Channels*, Chichester, UK: Research Studies Press, John Wiley; 1981.

Chapter 3

Fundamentals of fluid flow

This chapter reviews the fundamental principles of fluid flow and explains how this affects the convection cooling of electrical machines. For all fluid flow problems, the governing equations can be solved numerically based on the conservation of mass, momentum, and energy. Since the fluid flow equations are non-linear, they are being solved iteratively commonly using computational fluid dynamics (CFD) software. The main drawback of using the CFD method is its expensive computational cost and also the user needs to have good skills in CFD before useful solutions can be obtained, see Section 5.2.

Nevertheless, the relationship between the flow rate, fluid pressure, and the resistance to fluid flow can be simply explained by the flow equations given in this chapter. The flow rate passing through a cooling duct is limited by two main factors. One is the pressure gain from an impeller (fan/pump). Euler's turbo-machinery formula is a useful equation that provides a fundamental understanding of the key parameters that affect the pressure gain from an impeller while the fan affinity laws are a useful tool for estimating the relationships between pressure, flow rate, and impeller power with impeller size and speed. The other factor that affects the flow rate is the flow resistance due to duct wall friction and flow separation effects. Besides the pressure losses in stationary ducts, some cooling ducts such as the annular air gap and rotor ducts suffer additional rotational pressure losses due to the effects of Coriolis force and centrifugal force. This means that for a constant fan pressure, the amount of flow passing through the air gap and rotor ducts is less than the stationary condition. Many correlations that describe the variation of pressure loss coefficient with the rotational speed based on the experimental investigation have been provided in this chapter. These correlations are useful tools that allow electrical machine designers to calculate the convective cooling performance.

3.1 Basic principles of fluid flow

The cooling of electrical machines depends on two approaches: passive cooling and active cooling. Passive cooling refers to the effectiveness of heat spreading within a machine and is affected by material thermal properties, geometrical design, and interfacial thermal resistances between the machine components. For example, the passive thermal design of a stator slot strongly influences the temperature gradient within the coil. Active cooling refers to heat removal from a machine to the coolant based on forced convection of the cooling fluid. Active convection cooling involves fluid flow, which is described in this chapter.

In fluid mechanics, the Bernoulli theorem is the basic principle that describes the relationship between flow velocity, pressure, and elevation in an ideal fluid flow, based upon the conservation of energy. It can be written as:

$$p_1 + \frac{1}{2}\rho V_1^2 + \rho g z_1 = p_2 + \frac{1}{2}\rho V_2^2 + \rho g z_2 \qquad (3.1)$$

where p is the pressure energy, $\frac{1}{2}\rho V^2$ is the kinetic energy, and $\rho g z$ is the potential energy. They remain constant because the Bernoulli's principle is only applicable for inviscid flow. Cooling media used for electrical machines are real fluids that have viscosity. Viscous flow results in energy loss due to frictional effects and flow separation. The frictional affect is associated with shear stresses in the fluid as a result of viscosity. Flow separation losses arise from flow disturbances due to changes in flow cross-sectional area, changes in the flow direction, and interaction with other pipe fittings such as bends, valves, flow measurement devices, and junctions. This energy loss is commonly known as pressure loss or pressure drop. With the presence of pressure losses (p), the Bernoulli's equation can be rewritten as:

$$(p_1 - p_2) + \frac{1}{2}\rho\left(V_1^2 - V_2^2\right) + \rho g(z_1 - z_2) = \Delta p \qquad (3.2)$$

Consequently, a fan or pump is required to overcome the pressure drop where the fan or pump generates a differential pressure to maintain the flow. As active convective cooling is strongly affected by the motion of the fluid, machine designers need to pay attention to these topics as explained further in later sections.

3.1.1   Viscosity and boundary layer

The viscosity of a fluid is a measure of its resistance to gradual deformation by shear stress. When a fluid flows over a static surface, the flow velocity in the x-direction, parallel to the boundary, decreases from its maximum value to zero from the free stream towards the solid boundary, as shown in Figure 3.1. The distance between the free stream and the boundary surface is known as the boundary layer thickness, δ. The shear resistance is caused by internal friction in the fluid between the molecules when layers of fluid layer slide over one another. The shear stress, that opposes the relative motion of these layers, can be represented as:

$$\tau = \mu \frac{du}{dy} \qquad (3.3)$$

where μ is the dynamic viscosity which indicates the shear stress generated by a velocity gradient in a fluid. Viscosity is strongly influenced by temperature; therefore, temperature effects should not be neglected.

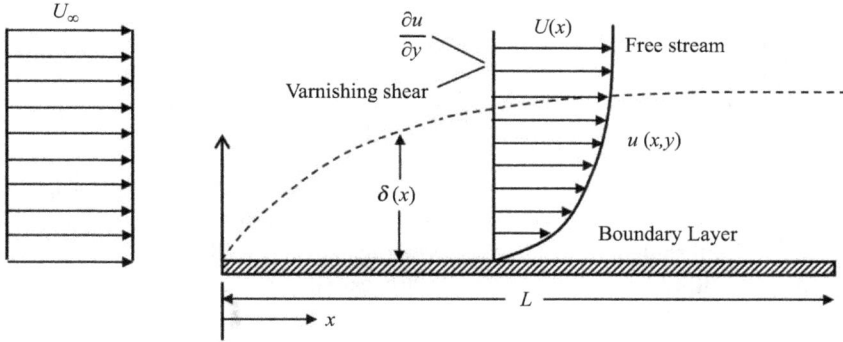

Figure 3.1   Boundary layer flow

3.1.2   Navier–Stokes equations

Fluid motion is governed by conservation equations based upon the basic physical laws of fluid mechanics. These are as follows:

The conservation of mass – the rate of change of mass in a fluid element is equal to the net flow rate of mass into the fluid element which is given by:

$$\frac{\partial \rho}{\partial t} + \nabla \cdot \rho V = 0 \tag{3.4}$$

Equation (3.4) is also known as the continuity equation. V is a velocity vector.

The conservation of momentum – the rate of change of momentum of a fluid particle is equal to the sum of the forces acting on the fluid element. This is known as the Navier–Stokes equation. The terms ∇p, $\nabla \cdot \tau_{ij}$, and ρg on the right-hand side of (3.5) are pressure, viscous, and gravitational force components, respectively. The viscous force is due to viscous stress components acting on a fluid element. For turbulent flow, there are additional stresses due to the viscous stress term, known as Reynolds stresses. These Reynolds stresses due to turbulence have been studied by many researchers and led to different turbulence modeling methods in the numerical analysis of fluid flow:

$$\rho \left(\frac{\partial V}{\partial t} + V \cdot \nabla V \right) = -\nabla p + \nabla \cdot \tau_{ij} + \rho g \tag{3.5}$$

The conservation of energy – the rate of change of energy of a fluid element is equal to the sum of the net rate of heat added to the fluid element and the net rate of work done on the fluid element:

$$\frac{\partial T}{\partial t} + V \cdot \nabla T = \alpha \nabla^2 T + \Phi \tag{3.6}$$

where α is the thermal diffusivity of a fluid.

The governing equations, presented in the differential form, are derived by considering an infinitesimal fixed control volume. As these conservation equations are nonlinear partial differential equations, the conjugate heat transfer problem of electrical machines can also be solved numerically using the CFD method.

3.1.2.1   Turbulent flow and modeling

Equations (3.4) – (3.6) are the governing equations that can be solved for laminar flow. However, electrical machines often operate in a turbulent flow regime. CFD simulation for turbulent flow is more complicated due to the small-scale eddies present in a turbulent flow field. They are called turbulent eddies as they are unsteady, three-dimensionally random, chaotic, and swirling. The Reynolds number is a common measure of the flow regime in a flow. The definition will be explained in a later section, but a high Reynolds number leads to turbulent flow as the inertia forces are sufficiently large to trigger the turbulent eddies in a flow.

The Reynolds-Averaged Navier–Stokes (RANS) approach is widely used for modeling turbulent flow because in general, the effects of turbulence on the mean flow are usually sufficient to quantify the turbulent flow characteristics. To obtain the RANS equations, the velocity and all other variables in a turbulent flow field are decomposed into time-averaged and fluctuating components.

For the flow velocity,

$$
\begin{aligned}
u &= \bar{u} + u' \\
v &= \bar{v} + v' \\
w &= \bar{w} + w'
\end{aligned}
\tag{3.7}
$$

For the pressure and temperature,

$$
\begin{aligned}
p &= \bar{p} + p' \\
T &= \bar{T} + T'
\end{aligned}
\tag{3.8}
$$

The RANS approach adopts the time-averaging operation on the Navier–Stokes equations and discards all details concerning the state of flow contained in the instantaneous fluctuations, resulting in the equations with the mean quantities and an additional term in the momentum transport equation due to the fluctuations of flow velocity. The additional term is known as the Reynolds stress tensor and it can be defined as:

$$
\tau_{turb} = -\rho \overline{u_i' u_j'} = -\rho \begin{pmatrix} \overline{u'u'} & \overline{u'v'} & \overline{u'w'} \\ \overline{u'v'} & \overline{v'v'} & \overline{v'w'} \\ \overline{u'w'} & \overline{v'w'} & \overline{w'w'} \end{pmatrix}
\tag{3.9}
$$

Hence, the viscous force term of the momentum equation is the sum of viscous stresses of the working fluid and turbulent stresses (Reynolds stresses) as expressed in the following equation:

$$
\tau_{ij} = \mu \left(\frac{\partial u_i}{\partial x_j} + \frac{\partial u_j}{\partial x_i} \right) - \rho \overline{u_i' u_j'}
\tag{3.10}
$$

Since the Reynolds stress tensor is symmetric, it creates an additional six unknowns in the RANS approach. So, it is necessary to have the same number of closure equations to obtain a solution. To solve for the Reynolds stresses, eddy viscosity models are applied, which are based on the Boussinesq approximation to relate the Reynolds stresses to the mean rates of deformation:

$$-\rho\overline{u_i'u_j'} = 2\mu_t S_{ij} - \frac{2}{3}\rho k\delta_{ij}, \quad S_{ij} = \frac{\partial u_i}{\partial x_j} + \frac{\partial u_j}{\partial x_i} \tag{3.11}$$

where μ_t is the turbulent viscosity, S_{ij} is the mean strain rate tensor, and δ_{ij} is the Kronecker delta function (i.e., the function is 1 if the variables are equal, or otherwise is 0). k is the turbulent kinetic energy, which is expressed as:

$$k = \frac{1}{2}\left(\overline{u'2} + \overline{v'2} + \overline{w'2}\right) \tag{3.12}$$

Similar to the momentum equation, turbulent eddy motion also generates an extra turbulent transport term (i.e., turbulent heat flux) in the energy equation. Thus, the turbulent heat flux can be modeled similarly. By analogy, the turbulent transport of heat is taken to be proportional to the gradient of the mean value of the temperature as:

$$-\overline{u_i'T'} = \alpha_t \frac{\partial \overline{T}}{\partial x_j} \tag{3.13}$$

where α_t is the turbulent thermal diffusivity, $\alpha_t = \lambda_t/\rho c_p$. The turbulent viscosity μ_t and turbulent thermal conductivity λ_t mathematically take into account the enhanced mixing and diffusion due to the turbulent eddies and hence provide closure of the Reynolds stresses and turbulent heat flux. These are related through the turbulent Prandtl number Pr_t. Hence, using the ensemble-averaged variables, the governing equations (3.4), (3.5), and (3.6) can be simplified to:

Continuity:

$$\frac{\partial \rho}{\partial t} + \nabla \cdot \rho \overline{V} = 0 \tag{3.14}$$

Momentum:

$$\rho\left(\frac{\partial \overline{V}}{\partial t} + \overline{V} \cdot \nabla \overline{V}\right) = -\nabla \overline{p} + (\mu + \mu_t)\nabla^2 \overline{V} + \rho g \tag{3.15}$$

Energy:

$$\frac{\partial \overline{T}}{\partial t} + \overline{V} \cdot \nabla \overline{T} = (\alpha + \alpha_t)\nabla^2 \overline{T} + \Phi \tag{3.16}$$

There are a variety of turbulence models developed to compute the turbulent flow using the RANS approach. The RANS turbulence models rely on empirical constants that have been calibrated against comprehensive experiment data for a wide range of turbulent flows. Two-equation turbulence models are commonly used to model the turbulent flow of an electrical machine, e.g., $k-\varepsilon$ and $k-\omega$ turbulence models. The $k-\varepsilon$ turbulence model consists of a transport equation for the turbulent kinetic energy k and another equation to model the turbulence dissipation rate ε. While the $k-\omega$ turbulence model consists of a transport equation for the turbulent kinetic energy k and another equation to model the specific turbulence dissipation rate ω. It is important to note that based on the Boussinesq isotropic eddy viscosity assumption, two-equation turbulence models have limitations in modeling flow driven by anisotropic normal Reynolds stresses. As an alternative solution, the Reynolds stress transport model (RSM) is preferable as the directional effects of the Reynolds stress field are modeled in the exact Reynolds stress transport equations. The RSM is more computationally expensive than two-equation turbulence models as the six equations for Reynolds stress transport are solved along with a model equation for the dissipation rate of turbulent kinetic energy ε. To obtain higher-order CFD simulations for a turbulent flow, large eddy simulation (LES) or direct numerical simulation (DNS) can be used to resolve unsteady turbulent eddies but high computational resources are required.

3.1.3   Dimensional analysis and dimensionless parameters

The cooling of rotating electrical machines involves different disciplines such as fluid mechanics, thermodynamics, and heat transfer, frequently, referred to as thermofluids. Commonly, the equations to solve thermofluid problems are either not available or difficult to solve and experiments can be conducted. In thermofluid problems, dimensional analysis is a powerful and useful method involving grouping the variables involved into sets of dimensionless parameters. These can be used in a systematic empirical approach to allow the investigation of one arrangement to be applied to other situations having geometric and dynamic similarities. The dimensionless parameters commonly used in the cooling of rotational electrical machines are Reynolds number, Prandtl number, Nusselt number, and Grashof number.

Prandtl number (Pr) is a measure of the ratio of momentum diffusivity to thermal diffusivity (or the relative thickness of the thermal boundary layer versus the velocity boundary layer) and is defined as:

$$\mathrm{Pr} = \frac{\mu c_p}{\lambda} \tag{3.17}$$

where μ, λ, and c_p are the dynamic viscosity, thermal conductivity, and specific heat of the fluid, respectively. The Prandtl number of gases is normally less than 1. For example, the Pr of air and hydrogen at room temperature is ~0.7. However, the Prandtl number of liquids is normally greater than 1. For example, the Pr of water and standard automotive antifreeze fluid at room temperature is ~7 and ~39, respectively. This means liquids have thinner thermal boundary layers than gases. Therefore, heat transfer is easier for a given Reynolds number.

Reynolds number (Re) is used to quantify whether flow conditions lead to laminar or turbulent flow. It is a measure of the ratio of inertial forces to viscous forces, which is defined for duct flow as:

$$\text{Re} = \frac{\rho V d}{\mu} \tag{3.18}$$

where V is the mean fluid velocity, d is the diameter of a pipe, and ρ is the density of a fluid. Figure 3.2 illustrates the laminar and turbulent flow profiles in a pipe for a given flow velocity. The laminar flow profile is parabolic and the maximum flow velocity at the pipe centre is twice the mean flow velocity. For turbulent flow, the velocity profile is flatter in the central part of the pipe and the local flow velocity close to the pipe wall drops rapidly. The turbulent flow profile is commonly approximated by the one-seventh power-law velocity profile.

For laminar flow, each fluid particle moves along a streamline parallel to the pipe wall and consequently, heat can only be transferred by means of heat conduction from layer to layer. When the Reynolds number is large enough to trigger lateral mixing of the fluid particles, the flow becomes turbulent which is more favourable for heat transfer. For internal duct flows, a flow remains laminar up to Re of 2,300 and becomes fully turbulent when Re is greater than 5,000. However, in practice, the flow may not be fully turbulent until Re is 10,000. The flow transition from laminar to turbulent occurs for Re between 2,300 and 10,000. Consequently, convective heat transfer strongly depends on values of the dimensionless numbers above as depicted in Figure 3.3.

Based upon the flow regime and the fluid type, the convective heat transfer coefficient is characterized by the Nusselt number (Nu). Nusselt number is a measure of the ratio of convective to conductive heat transfer normal to a boundary.

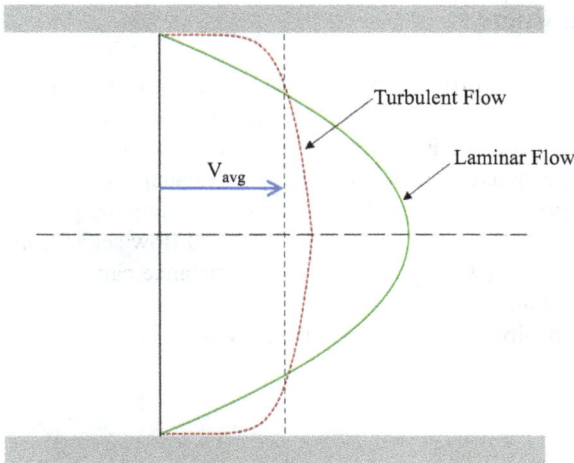

Figure 3.2   Laminar and turbulent flow profiles in a pipe for a given flow velocity

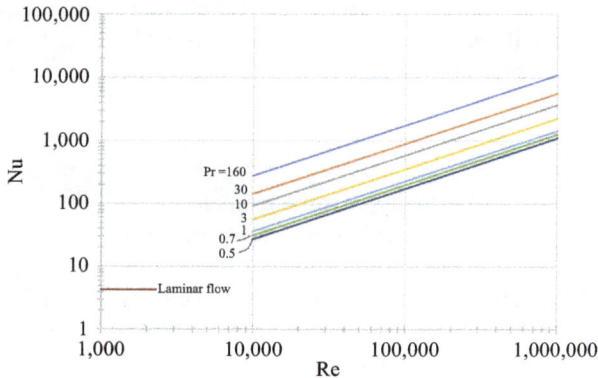

Figure 3.3   The variation of Nusselt number with Reynolds number and Prandtl number for fully developed velocity and temperature profiles, circular duct, and constant heat flux

When Re is greater than 10,000, the turbulent flow leads to higher convective heat transfer. This is further enhanced by liquid coolants of a higher Prandtl number. Based upon the empirical method, Nusselt number for turbulent flow is normally a function of the Reynolds number and Prandtl number which is expressed as:

$$\text{Nu} = f(\text{Re}, \text{Pr}) \tag{3.19}$$

Nusselt number correlations for various geometries are widely available and a detailed explanation of convective heat transfer calculation methods has been provided in Chapter 2.

3.2   Flow in ducts

As explained earlier, a differential pressure is required to produce flow in a duct. The pressure created by a fan or pump is used to overcome the pressure loss in the flow system. The flow rate is determined from the intersection between the fan (or pump for liquid) characteristic and system flow resistance curves. At the "operating point," the pumping pressure produced by the fan meets the pressure requirements of the system at that flow rate due to frictional and flow separation pressure losses as described in Section 3.1. The system flow resistance can be represented by R as shown in the formulations below. The system pressure loss (Δp) increases in a parabolic relationship with the volumetric flow rate (Q):

$$\Delta p = RQ^2 \tag{3.20}$$

$$R = \frac{K\rho}{2A^2} \tag{3.21}$$

where K is the pressure loss coefficient, ρ is the fluid density, and A is the flow cross-section area. Figure 3.4 illustrates a typical fan characteristic curve.

The fan pressure rise is not constant and reduces with the flow rate passing through the fan due to energy losses. A reduced flow resistance can give a higher flow rate whereas increased flow resistance can reduce the flow rate in a system. Besides the traditional frictional and flow separation losses, the operation of electrical machines involves rotation that will cause additional flow resistance in the system. As a result, for the same fan pressure, the flow rate through the system will be less than that of the stationary condition. The additional flow resistances due to rotation are presented in the sections below.

3.2.1 Pressure loss

Two types of pressure loss need to be overcome by a fluid: frictional loss and flow separation loss. The frictional pressure loss is due to the fluid friction on account of viscosity. The Darcy–Weisbach equation describes the relationship between friction factor (f), duct length (L), duct diameter (d), and the flow kinetic energy:

$$\Delta p = f\frac{L}{d} \times \frac{1}{2}\rho V^2 \tag{3.22}$$

The combination of friction factor, duct length, and duct diameter forms the frictional loss coefficient which is expressed as:

$$K = f\frac{L}{d} \tag{3.23}$$

where f is the Darcy friction factor. It is affected by flow regime (laminar and turbulent flow) and duct wall surface roughness. In Figure 3.5, the Moody chart shows that laminar flow is independent of surface roughness while turbulent flow is surface roughness dependent. The effect of surface roughness is more dominant at a higher Reynolds number.

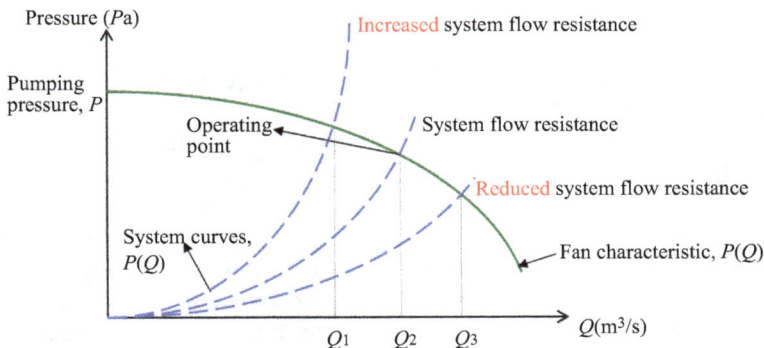

Figure 3.4   Fan characteristic curve vs system flow resistance curve

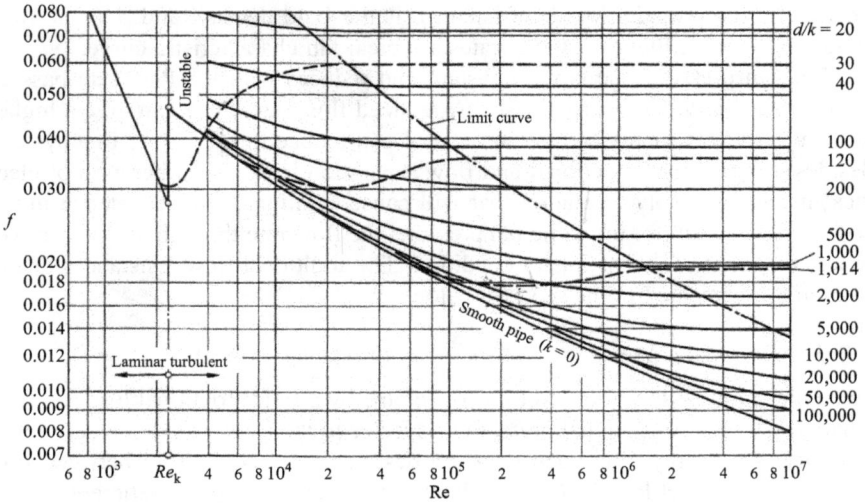

Figure 3.5   Moody chart [1]

As the frictional loss is proportional to the square of the flow velocity in the duct, the frictional pressure loss can be reduced by increasing the flow cross-section area. By referring to (3.20) and (3.21), for example, doubling the flow cross-sectional area means a 75% reduction in pressure loss for a given flow rate. This can be achieved by increasing the duct diameter or number of ducts. Due to space limitations, machine design engineers commonly increase the number of parallel flow paths to reduce pressure loss. As illustrated in Figure 3.6, the number of rotor ducts in an induction machine is doubled to meet the fan characteristic curve.

Often, the cross-section of cooling ducts is non-circular. For a non-circular duct, the hydraulic diameter (D_h) can be applied which is defined as:

$$D_h = \frac{4A}{P'} \tag{3.24}$$

where A is the cross-sectional area and P' is the wetted perimeter of the non-circular duct.

Although the Darcy friction factor can be determined easily from the Moody chart, the friction factor obtained only accounts for fully developed flow. It does not account for developing flow, which is the case in most electrical machines mainly due to the change in flow cross-section area. As shown in Figure 3.7, the friction loss in the entrance region is much higher than the friction loss for fully developed flow. After the entrance region, the friction loss varies linearly with the duct length. Therefore, the friction factor needs to be corrected for developing flow for a more accurate calculation of the frictional pressure loss.

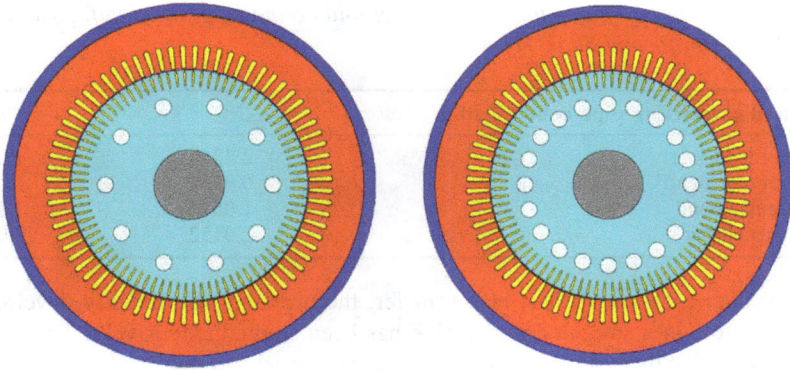

Figure 3.6   The flow cross-section of an induction machine is doubled by increasing the number of rotor ducts from 10 (left) to 20 (right)

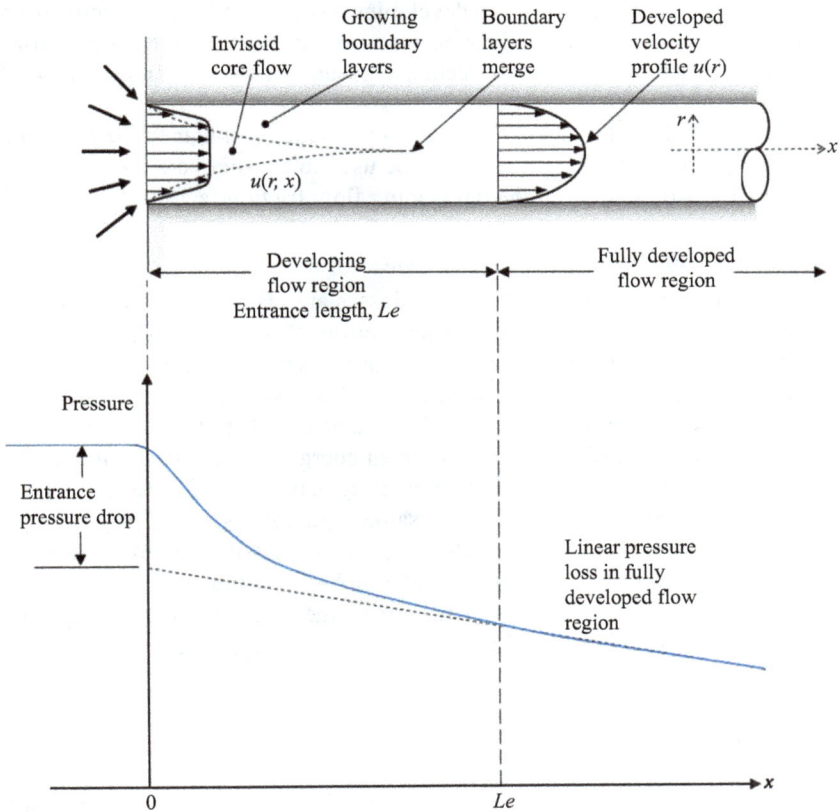

Figure 3.7   Developing velocity profiles and pressure changes in the entrance of a duct

Table 3.1   C and n values for various inlet and connecting pipe configurations, Bhatti and Shah [2]

Inlet and connecting pipe configuration	C	n
Square-edged inlet	2.4254	0.676
180° circular bend	0.9759	0.700
90° circular bend	1.0517	0.629
90° sharp elbow	2.0152	0.614

For duct flow convection heat transfer, the problem of thermally developing turbulent flow in a smooth circular pipe has been studied extensively. To account for the impact of thermal developing flow, an empirical formula can be used for air:

$$\frac{\text{Nu}_m}{\text{Nu}_\infty} = 1 + \frac{C}{(L/D_h)^n} \tag{3.25}$$

where Nu_∞ is the fully developed turbulent flow Nusselt number and Nu_m is the mean Nusselt number for thermally developing flow. C and n are empirical constants that vary with duct inlet and connecting pipe configurations before entering a duct where convective heat transfer occurs, as given in Table 3.1. In general, (3.25) is valid for $L/D_h > 3$.

For air ($\text{Pr} = 0.7$), the developing velocity profile is similar to the developing temperature profile. Therefore, (3.25) can be used to approximate the influence of duct entrance configuration on the developing flow friction factor.

3.2.2   Flow separation pressure loss

Apart from friction loss, all other pressure losses are classified as flow separation pressure loss. Flow separation occurs whenever the change in velocity of the fluid in either magnitude or direction, causes boundary layer separation. Due to adverse pressure gradients, the fluid particles are forced to move away from the surface resulting in flow separation. This flow phenomenon leads to the formation of turbulent eddies and is associated with significant energy loss. In electrical machines, flow separation pressure losses in the cooling paths are due to contractions, expansions, or bends in the flow. The same equation as the Darcy–Weisbach equation can be used to calculate the flow separation pressure loss by replacing the friction loss coefficient with the flow separation loss coefficient.

As flow separation loss is associated with the effect of turbulence, a flow separation loss is usually quantified with an empirical loss coefficient (K) based upon the flow kinetic energy as:

$$\Delta p = K \times \frac{1}{2} \rho \, V^2 \tag{3.26}$$

Information on pressure loss coefficients for different flow separation cases is available in Idelchik [3] and Miller [4]. Some common loss coefficients are given in Table 3.2.

Table 3.2 Loss coefficient of common flow components (A_i and A_o are inlet and outlet cross-section areas, respectively, R/D_h is the relative radius of curvature)

Flow component	K-value
Sudden expansion flow	$K = \left(1 - \frac{A_i}{A_o}\right)^2$
Sudden contraction flow	$K = 0.5\left(1 - \frac{A_o}{A_i}\right)^{0.75}$
90° bending flow with a circular or square cross-section	when $0.5 < \frac{R}{D_h} < 1$ $K = 0.21\left(\frac{R}{D_h}\right)^{-2.5}$ when $\frac{R}{D_h} \geq 1$, $K = 0.21\left(\frac{R}{D_h}\right)^{-0.5}$
45° sharp elbow flow with a circular or square cross-section	$K = 0.32$
90° sharp elbow flow with a circular or square cross-section	$K = 1.2$

For open ventilated cooling machines, a grill or filter is often added in the flow loop to prevent blockage of the cooling passages due to dust contamination. However, the usage of a grill or filter will cause additional pressure loss. The characteristic shown in Figure 3.8(a) is used to calculate pressure drop at the entry to the system due to a grill or filter over the inlet vents. A similar characteristic is used for outlet vents, as shown in Figure 3.8(b). Both use a combination of data from Osborne and Turner [6] and Lightband and Bicknell [7]. The loss coefficient is a function of the free area ratio. For both inlet and outlet, the flow area used in the flow resistance calculation is referring to the area with a smaller value. Hence, flow area after inlet grill is used for inlet whereas flow area before outlet grill is used for an outlet.

Figure 3.9 shows the pressure gradient in a flow in a circular pipe when a diameter of 30 mm contracts suddenly into a circular pipe to a diameter of 15 mm. This exhibits a substantial pressure loss due to the sudden contraction. Due to duct wall friction, the pressure drops linearly along the pipe before the contraction and after the flow becomes re-developed. The pressure loss coefficient can be then extracted from the pressure difference before and after the contraction and also the flow kinetic energy downstream.

3.3   Fans and rotor driven pressure gains

The flow of a fluid in ducts and the pressure loss due to friction and flow separation have already been discussed in the previous section. Since convective cooling is dependent on the flow rate through the cooling system, sufficient pressure needs to be provided from rotodynamic devices (impeller) to overcome the pressure loss. For electrical machines with active cooling, fans and pumps are the most common devices to provide energy to the fluid. Fans increase the pressure in gases, while pumps increase the pressure in liquids. The type of impeller is distinguished by the direction of flow in relation to the axis of rotation of the impeller, see Figure 3.10. *Axial flow* impeller creates the flow that is parallel to the impeller rotation axis. *Radial flow* impeller (centrifugal) creates the flow that is normal to the impeller rotation axis. The impeller that delivers both axial and radial flows is called a *mixed flow* impeller. Centrifugal fans and pumps are the most common type of impellers used in electrical machines.

For an air-cooling example, the totally enclosed fan-cooled (TEFC) machine has a shaft-mounted centrifugal fan at the non-drive end outside the machine. This draws ambient air axially into the impeller and the air is forced to flow over the outer housing fins (see Figure 3.11(a)). The totally enclosed self-ventilated machine has a shaft-mounted centrifugal fan inside the machine generating circulating air flow between the rotor ducts and stator ducts (see Figure 3.11(b)).

3.3.1   Euler's turbomachinery equation

In reality, the flow through the impeller is three-dimensional and therefore very complex depending on the impeller blade number, thickness, shape, blade width,

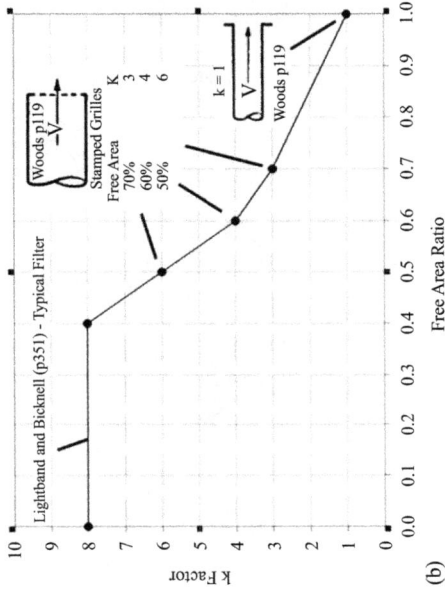

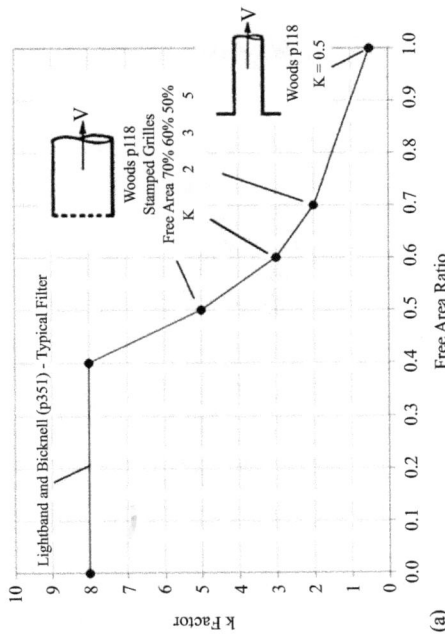

Figure 3.8 Loss coefficient of inlet and outlet grill: (a) inlet grill and (b) outlet grill [5]

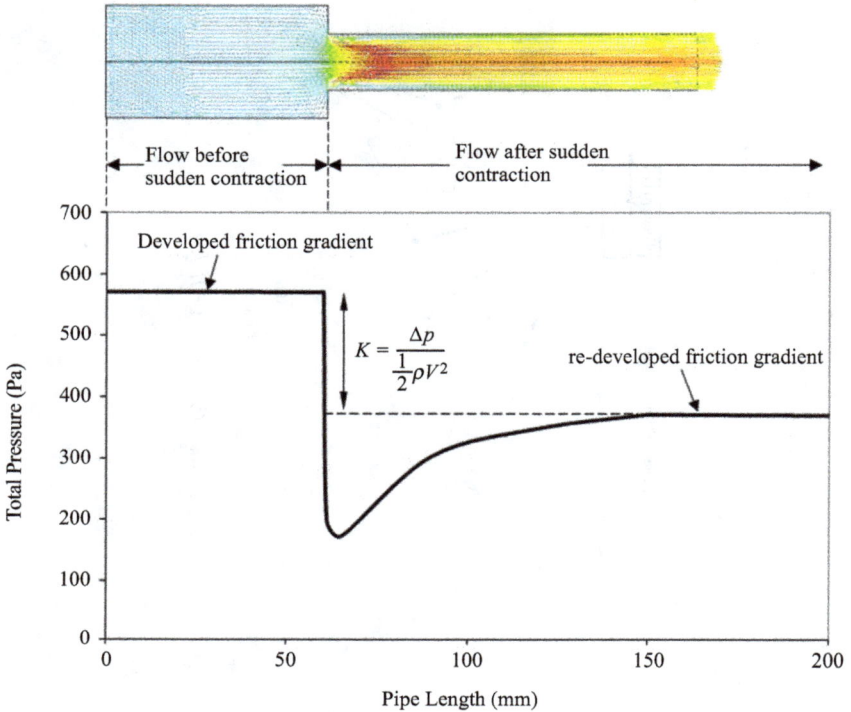

Figure 3.9   *Pressure gradient of a sudden contraction flow in a circular pipe of diameter 30 and 15 mm*

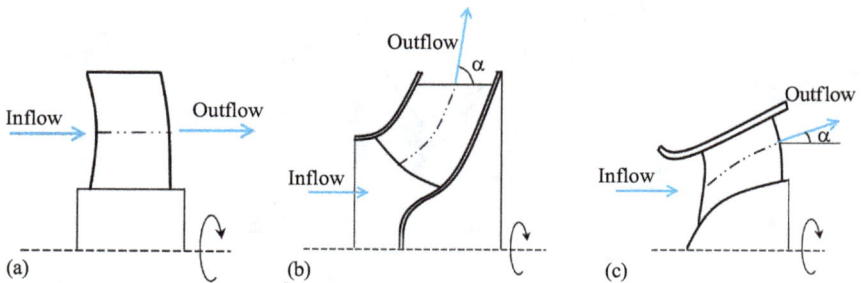

Figure 3.10   *Different impeller types: (a) axial flow impeller, (b) radial flow impeller, and (c) mixed flow impeller*

blade radius, and even its supporting hub. However, the flow can be simplified considerably by a one-dimensional theory that provides a good understanding of fundamental relationships between important parameters (impeller geometry, rotational speed, etc.) that affect the pressure gains in the fluid from the impeller. In one-dimensional flow analysis, the impeller section is represented by an

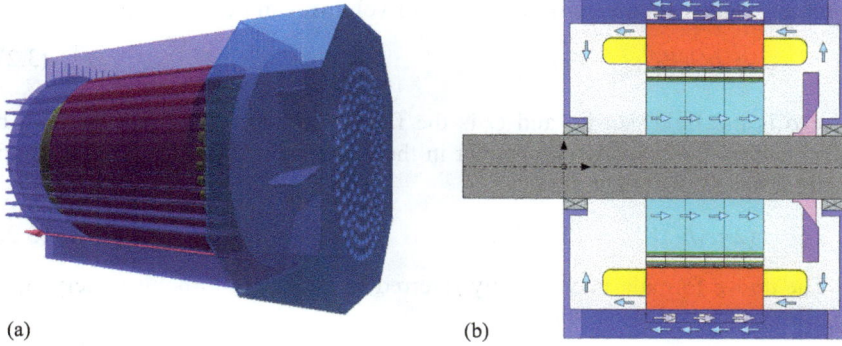

Figure 3.11   The usage of impellers in electrical machine cooling: (a) TEFC machine and (b) totally enclosed self-ventilated machine

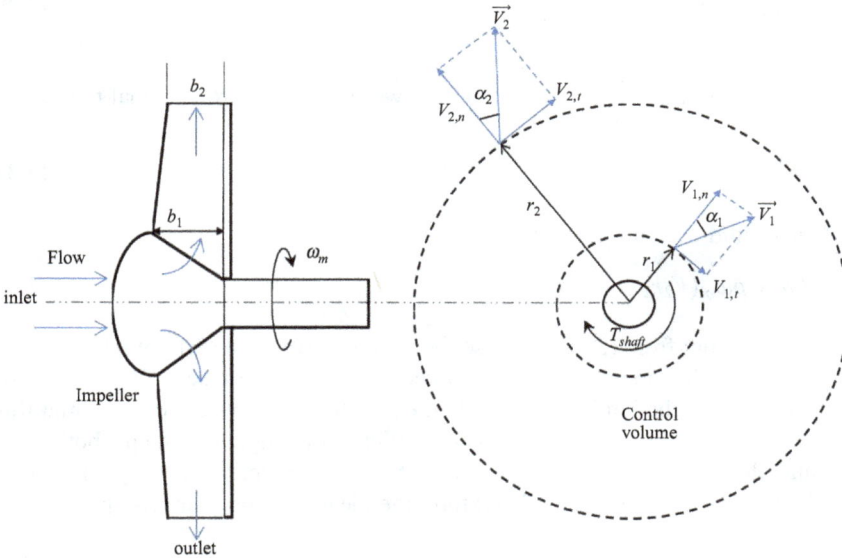

Figure 3.12   One-dimensional flow diagram of a centrifugal impeller

annular control volume as shown in Figure 3.12. The shaft rotates at an angular velocity of ω_m.

Flow enters the control volume with an absolute velocity of $\vec{V}_1$ at radius r_1 and exits with an absolute velocity of $\vec{V}_2$ at radius r_2. As the moment of momentum is defined as the cross product of $\vec{r} \times \vec{V}$, only the velocity component in the tangential direction ($V_{1,t}$ and $V_{2,t}$) contributes to the torque. Mechanical energy is applied to rotate the impeller and the shaft torque transmitted to the fluid, causing a rate of change of angular momentum in the control volume, can be written as:

$$T_{shaft} = \rho Q(V_{2,t}r_2 - V_{1,t}r_1) \tag{3.27}$$

where ρ is the fluid density and Q is the fluid volumetric flow rate that can be determined by the velocity component in the normal direction ($V_{1,n}$ and $V_{2,n}$) and the circumferential area as:

$$Q = (2\pi r_1 b_1)V_{1,n} = (2\pi r_2 b_2)V_{2,n} \tag{3.28}$$

The *above* formula is commonly referred to as Euler's turbomachinery equation. Then, the shaft power is:

$$P_{shaft} = T_{shaft} \times \omega_m = \rho\omega_m Q(V_{2,t}r_2 - V_{1,t}r_1) \tag{3.29}$$

In a fluid power system, the fluid power is expressed as:

$$P_{fluid} = \Delta p \times Q \tag{3.30}$$

where Δp is the fluid pressure rise.

Due to conservation of energy, fluid power is directly proportional to the shaft power as:

$$P_{fluid} \propto P_{shaft} \tag{3.31}$$

Thus, fluid pressure rise is:

$$\Delta p \propto \rho\omega_m(V_2 r_2 - V_1 r_1) \tag{3.32}$$

Q cancels out from (3.32) because the fluid pressure is mainly of interest and it is given from the operating point between fan characteristic and system flow resistance as described in Section 3.2. In the simplified analysis, the tangential fluid velocity is assumed to be equal to the impeller blade angular velocity both at the inlet and the outlet. The impeller blades have a tangential velocity of $\omega_m r_1$ at the inlet and $\omega_m r_2$ at the outlet. Therefore, the ideal fluid pressure rise provided by the impeller can be represented as:

$$\Delta p \propto \rho\omega_m{}^2(r_2^2 - r_1^2) \tag{3.33}$$

As all fans and pumps suffer from irreversible losses due to wall friction and flow separation on the blade surfaces, flow leakage, turbulent dissipation, etc., the energy that is delivered to the fluid is less than the shaft torque. Hence, the useful pressure rise supplied to the fluid is:

$$\Delta p = C\rho\omega_m^2(r_2^2 - r_1^2) \tag{3.34}$$

where C is the loss coefficient of the fan. It is normally obtained using empirical methods.

3.3.2   *Fan's laws*

The above section has shown the important parameters that affect the pressure gains from an impeller. Often electrical machine designers would like to estimate the effect of rotational speed and the change of impeller size on the fluid pressure rise delivered. Then, the method of dimensional analysis as described in Section 3.1.3 becomes very useful. For fan engineering, the dimensionless parameters involved are capacity coefficient, pressure coefficient, and power coefficient.

The capacity coefficient is a measure of the volumetric flow rate induced, defined as:

$$C_Q = \frac{Q}{\omega_m D^3} \tag{3.35}$$

The pressure coefficient is a measure of the fluid pressure gain, defined as:

$$C_H = \frac{p}{\rho \omega_m^2 D^2} \tag{3.36}$$

The power coefficient is a measure of the power consumption, defined as:

$$C_p = \frac{p_{shaft}}{\rho \omega_m^3 D^5} \tag{3.37}$$

It is important to note that D here refers to fan size. Though D commonly refers to the fan outer diameter, other dimensions must be scaled in the same way to meet the condition of geometric similarity. These dimensionless parameters are very useful for relating prototype fan performance to a model fan if they meet both geometric and dynamic similarities. The fan scaling laws derived are called the affinity laws and they are used to express the relationship of the ratios between fluid pressure, volumetric flow rate, rotational speed, fan size, and shaft power. For any two fans states 1 and 2,

$$\frac{Q_2}{Q_1} = \frac{\omega_{m2}}{\omega_{m1}} \left(\frac{D_2}{D_1}\right)^3 \tag{3.38}$$

$$\frac{\Delta p_2}{\Delta p_1} = \frac{\rho_2}{\rho_1} \left(\frac{\omega_{m2}}{\omega_{m1}}\right)^2 \left(\frac{D_2}{D_1}\right)^2 \tag{3.39}$$

$$\frac{p_2}{p_1} = \frac{\rho_2}{\rho_1} \left(\frac{\omega_{m2}}{\omega_{m1}}\right)^3 \left(\frac{D_2}{D_1}\right)^5 \tag{3.40}$$

Example: A designer needs to calculate the fluid pressure gains from a fan when the fan speed is doubled, while the fan is working with the same fluid, $\rho_1 = \rho_2$ and the same fan size $D_1 = D_2$. Equation (3.39) can be then applied here and the fluid pressure at doubled speed is approximately 4 times the original fluid pressure as demonstrated below:

$$\Delta p_2 = \left(\frac{2 \times \omega_{m1}}{\omega_{m1}}\right)^2 \Delta p_1 = 4\Delta p_1 \tag{3.41}$$

The fan scaling laws are a useful design tool for a designer to estimate the scaled fan performance at different conditions if the performance curve of an existing fan is known. However, the scaling laws are less representative when the size ratio of the full-scale prototype to the small-scale model is large. The prototype performance is generally better because commonly it is not necessary for all the geometric dimensions to be increased by the same scaling factor. For example, the full-scale prototype could have a smaller tip clearance relative to the fan blade outer diameter which gives less flow leakage and tip loss. The relative roughness, which is a measure of surface roughness to fan blade diameter ratio, may also be much smaller due to the bigger fan size, which gives less wall friction loss.

3.4   Flow in the air gap

For electrical machines, the flow in the air gap between stator and rotor is highly sheared due to rotation and it becomes unstable with the formation of complex toroidal vortices. This shear flow is commonly known as Taylor vortex flow [8]. The formation of Taylor vortices is measured by the Taylor number (Ta) and the Taylor vortices are formed in the air gap if Ta is greater than a critical value, i.e., ≈ 41 for a narrow annular gap. The Taylor number (Ta) is defined as:

$$\text{Ta} = \frac{\rho \omega_m \, r_m^{0.5} s^{1.5}}{\mu} \tag{3.42}$$

where s is the air gap size, r_m is the mean rotor–stator radius, i.e., $r_m = (r_i + r_o)/2$, r_i and r_o are inner and outer radii of the air gap.

Above the critical value, the viscous forces that dominate in Couette flow are overcome by the centrifugal forces. A secondary flow develops in the form of regularly spaced toroidal vortices, rotating in opposite directions along the annulus, as illustrated in Figure 3.13. These Taylor vortices can significantly enhance the convective heat transfer within the air gap when compared to the case without Taylor vortices. Becker and Kaye [10] have proposed suitable correlations for air gap convective heat transfer prediction.

For through ventilated machines, the air gap flow has a superimposed axial flow and this axial flow can provide additional convective heat transfer in the air gap. However, at the same time, the superimposed axial flow can suppress and reduce the formation of Taylor vortices when compared to the case without the axial flow. Kaye and Elgar [11] revealed the existence of four possible modes in the

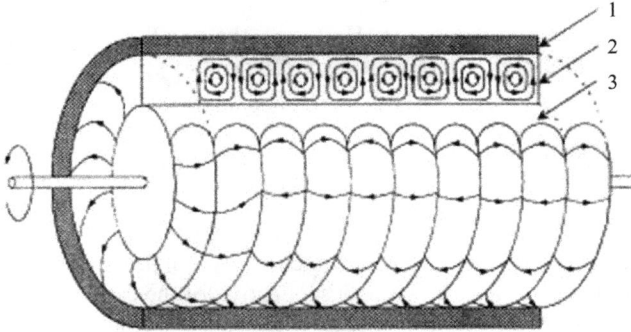

1 : stationary outer cylinder

2 : Taylor vortices

3 : rotating inner cylinder

Figure 3.13   Taylor vortices [9]

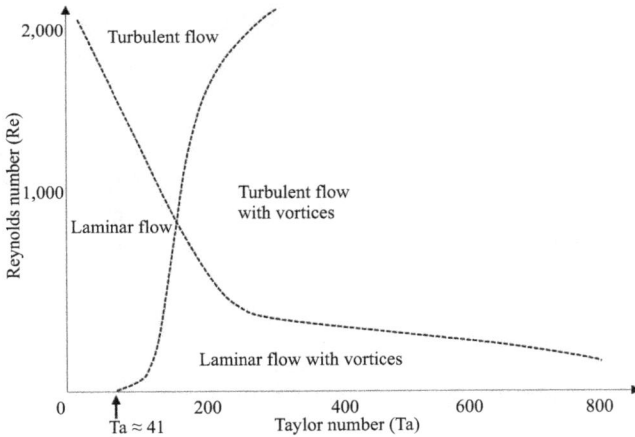

Figure 3.14   Modes of mixed axial and shear flow in an annulus between concentric cylinders, (radius ratio, $\eta \approx 1$) [11]

air gap. These are categorized by the Taylor number and axial Reynolds number as illustrated in Figure 3.14 as follows:

1.   Pure laminar flow
2.   Laminar flow with Taylor vortices
3.   Turbulent flow
4.   Turbulent flow with Taylor vortices

Convective heat transfer correlations have been proposed by many researchers, which depend on the type of flow in the air gap (more details in Chapter 2). To calculate the flow type and its air gap heat transfer, the challenge is to determine the axial flow velocity in advance. However, the axial flow is subjected to the effects of rotation, and the flow resistance for flow to pass through the air gap is higher.

3.4.1   Rotational pressure losses in the air gap

The variation of inlet pressure over a range of air gap flow rates for rotor speeds up to 3,000 rpm is shown in Figure 3.15, by Chong [12]. This demonstrates the air gap flow suffers additional flow resistance which gives a higher pressure loss than the stationary case for a given flow rate. Consequently, for the case of rotation, the total pressure loss of the air gap can be defined as the sum of the stationary pressure loss (Δp_s) and rotational pressure loss (Δp_r):

$$\Delta p = \Delta p_s + \Delta p_r \tag{3.43}$$

Experimentally, the rotational pressure loss can be determined from the difference between the rotating and stationary conditions by assuming the stationary loss is the same amount in the rotating condition. Thus, the rotational loss is the excess pressure drop for a given flow rate. The rotational losses that arise are due to additional frictional loss, entrance loss at the air gap entry, and combining flow loss at the air gap exit.

3.4.2   Frictional pressure loss

With rotation, the flow in a rotor–stator gap becomes helical. Therefore, the length of the pathway travelled by the rotational flow is longer and thus this increases the friction pressure loss. The flow helix is strongly affected by the rotor speed and also the flow rate through the air gap, which can be characterized using rotation ratio (V_T/U). It is a measure of outer rotor tangential velocity to axial flow velocity. The friction factor of air flow passing through an air gap with an inner cylinder rotating

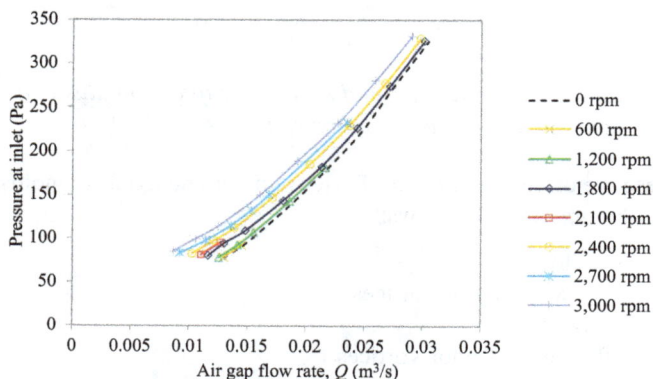

Figure 3.15   The variation of measured inlet pressure with the air gap flow rate

for turbulent flow can be estimated using the correlation proposed by Yamada [13]:

$$f_r = f_0 \left[1 + \left(\frac{7}{16} \right)^2 \left(\frac{V_T}{U} \right)^2 \right]^{0.38}$$

(3.44)

f_r is the friction factor for rotation and f_0 is the friction factor for stationary, i.e., Darcy's friction factor.

3.4.3 Entrance pressure loss

The operation of electrical machines involves rotation which introduces additional forces to the system due to Coriolis and centrifugal accelerations. Therefore, a flow is subjected to the effect of the Coriolis force in a rotating reference frame. The Coriolis force is directly proportional to the mass of the fluid particle (m), the angular velocity vector of the rotating frame (ω_m), and also the velocity vector of the fluid particle (V) with respect to the rotating frame. The vector equation for the Coriolis force ($F_{Coriolis}$) can be expressed as follows:

$$F_{Coriolis} = -2m(\omega_m \times V)$$

(3.45)

The cross product operator (i.e., $\times$) in (3.45) multiplies the orthogonal ω and V vectors. The ω vector is directed along the axis of the rotating reference frame. It is important to note that the resulting Coriolis force is also orthogonal to both ω and V. The fluid around a rotor–stator gap is subjected to a pressure gradient and tends to flow towards the gap. At the same time, the fluid adjacent to the rotor is dragged by the rotor due to the no-slip condition. Therefore, in the rotating reference frame, the flow entering rotor–stator gap is deflected from its initial direction due to the Coriolis effect which is represented by red arrows in Figure 3.16.

To determine the entrance loss at the rotor–stator gap, an experimental investigation was conducted by Chong [12]. Dimensional analysis was performed to analyse the important parameters that affect the entrance loss. Based upon that analysis, the coefficient of entrance pressure loss at the rotor–stator gap entry can be correlated with the rotation ratio (V_T/U) as:

$$\begin{aligned} &\text{For } V_T/U > 1 \\ &K_{en} = 0.12(V_T/U)^2 - 0.12(V_T/U) \\ &\text{For } V_T/U \leq 1, \\ &K_{en} = 0 \end{aligned}$$

(3.46)

where V_T is the peripheral speed of the rotor outer surface and U is the mean axial flow velocity in the air gap. Also, the experimental investigation demonstrates that the entrance loss is negligible when the rotation ratio is less than unity. Consequently, Coriolis and centrifugal effects increase the pressure loss of fluid entering the rotor–stator gap. The additional pressure losses due to rotation need to be characterized before the convective heat transfer can be estimated accurately.

Moreover, to investigate the impact of gap size on the entrance pressure loss, CFD models were generated, compared, and calibrated against the experimental test data. As depicted in Figure 3.17, the analysis demonstrates that the entrance pressure loss is independent of the gap ratio, i.e., $G = s/r_i$. Hence, the proposed correlation (3.46) is applicable for most electrical machines of different gap size. Both CFD and experimental results show that the entrance loss is mainly affected by the rotation ratio and this needs to be taken into account for accurate flow modeling.

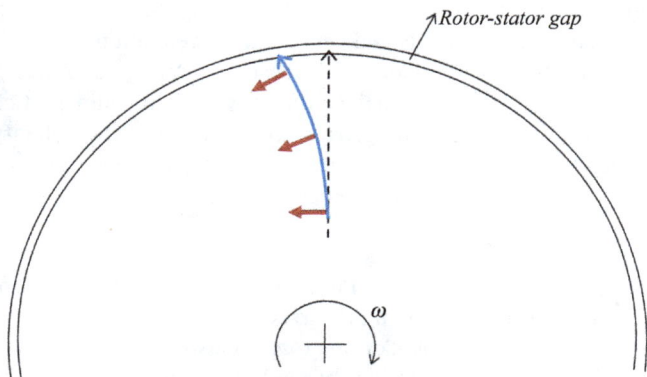

Figure 3.16   *The flow pattern at the entrance of rotor–stator gap in the rotating reference frame is represented by blue arrows. The dashed arrows represent the flow direction due to the pressure gradient force. The red arrows indicate the direction of Coriolis force acting on the fluid particles*

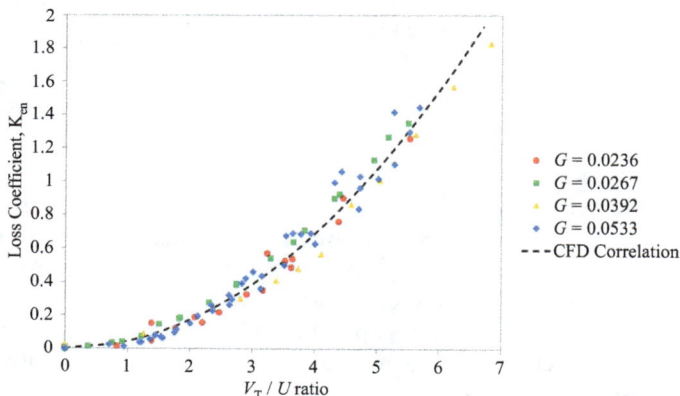

Figure 3.17   *The variation of entrance loss coefficient with rotation ratio over different gap ratios from 0.0236 to 0.0533*

3.5 Flow in rotating ducts

For certain electrical machines that need additional cooling for the rotor, circular ducts are often created in the rotor for convective air cooling. Similar to the air gap, the flow that passes through the rotor ducts, which rotate about the rotor axis parallel to the axes of the ducts, is also subjected to the Coriolis and centrifugal effects. The Coriolis force results in a swirling flow in the rotating duct entrance region as shown in Figure 3.18. Due to the decrease in the fluid velocity vector that is orthogonal to the axis of rotation, the Coriolis-induced secondary flow is predominant in the duct entrance region and decays away as the flow becomes fully developed downstream. Inside the duct, the flow tends to be parallel to the rotational axis. Then, the centrifugal effect becomes more important due to the fluid density difference created by heating at the duct wall. The centrifugal force acts outwards in the radial direction. The centrifugal force ($F_{centrifugal}$) is directly proportional to the square of angular velocity (ω_m), the mass of the fluid particle (m), and the distance of the fluid particle from the axis of the rotating reference frame (r) which can be written as:

$$F_{centrifugal} = m\omega_m^2 r \qquad (3.47)$$

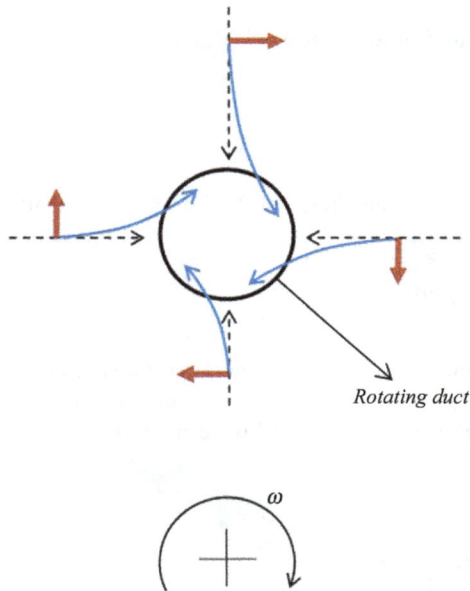

Figure 3.18 *The flow pattern at the duct entrance in the rotating reference frame is represented by blue arrows. The dashed arrows represent the flow direction due to the pressure gradient force. The red arrows indicate the direction of Coriolis force acting on the fluid particles*

The additional pressure losses due to rotation in the rotor ducts are the friction loss and the entrance loss.

3.5.1   Frictional pressure loss

As rotation induces secondary flows at the entrance of the ducts due to the Coriolis effect, this increases the resistance to the flow passing through ducts and hence gives a high-pressure loss. The increase in the friction factor of adiabatic flow under the effects of rotation has been investigated experimentally by Johnson and Morris [14]. As the rotation increases the flow resistance above the normal stationary condition, the correction factors (C) were obtained in terms of the ratio of the rotating case to the stationary case for given duct geometries.

For duct length-to-diameter ratio of 10.6 and laminar-like flow (i.e., $900 < $ Re $ < 9,880$):

$$C = \frac{f_r}{f_0} = 0.503J^{0.16}\mathrm{Re}^{-0.03} \tag{3.48}$$

For duct length-to-diameter ratio of 10.6 and turbulent-like flow (i.e., Re $\geq$ 9,880):

$$C = \frac{f_r}{f_0} = 0.842J^{0.023}\mathrm{Re}^{0.002} \tag{3.49}$$

For duct length-to-diameter ratio of 31.8 and laminar-like flow (i.e., $900 < $ Re $ < 7,000$):

$$C = \frac{f_r}{f_0} = 0.312J^{0.21}\mathrm{Re}^{0.01} \tag{3.50}$$

For duct length-to-diameter ratio of 31.8 and turbulent-like flow (i.e., Re $\geq$ 7,000):

$$C = \frac{f_r}{f_0} = 0.783J^{0.058}\mathrm{Re}^{-0.01} \tag{3.51}$$

where f_r is the rotational friction factor and f_0 is the stationary friction factor. J is the rotational Reynolds number and is a measure of the relative strength of the Coriolis forces to the viscous forces and is defined as:

$$J = \frac{\rho\omega_m d^2}{\mu} \tag{3.52}$$

3.5.2   Entrance pressure loss

Similar to the rotor–stator gap, the entrance pressure loss of rotating ducts was also investigated experimentally by Chong [12]. The investigation was performed by using a rotor with a series of straight circular ducts, which were parallel to the axis of the rotor. Based upon the additional pressure losses suffered by the air flow passing through the rotating ducts, the influence of duct size (d), duct pitch-circular radius of rotor ducts (H), and duct spacing (w) was examined. Based on the dimensional analysis, the entrance loss is found to increase considerably with the increase in rotation ratio (V_T/U), where V_T is the tangential speed of the rotor ducts rotating about a parallel axis and U is the mean axial flow velocity in the rotor ducts. But, it can be neglected when the rotation ratio is less than 0.5 approximately. Furthermore, the experimental results demonstrate that the entrance loss of the rotor ducts is independent of the duct size, duct length-to-diameter ratio (L/d), eccentricity parameter (H/d), and duct spacing ratio (w/d). However, the entrance loss is clearly affected by the proximity of the rotor ducts to the rotor periphery (H/r_i).

Consequently, for H/r_i ratio of 0.75, the entrance loss coefficient can be correlated with the rotation ratio as:

$$
\begin{aligned}
V_T/U > 0.5, \quad & K_{en} = 0.234(V_T/U)^2 - 0.043(V_T/U) \\
V_T/U \leq 0.5, \quad & K_{en} = 0
\end{aligned}
\tag{3.53}
$$

For H/r_i ratio of 0.5, the entrance loss coefficient can be correlated with the rotation ratio as:

$$
\begin{aligned}
V_T/U > 0.5, \quad & K_{en} = 0.474(V_T/U)^2 - 0.156(V_T/U) \\
V_T/U \leq 0.5, \quad & K_{en} = 0
\end{aligned}
\tag{3.54}
$$

3.5.3   Flow exits from rotating ducts

In open ventilated machines (see Chapter 4), it is very common for cooling ducts to be placed in the rotor, in parallel with the air gap. Inside the rotor ducts, the flow passing through the rotor ducts is restricted to rotate about the axis of the rotor. Due to centrifugal effects, the flow exiting from the rotating ducts moves radially outwards. It then hits the outlet flow from the air gap. The combining of the flow from the rotor ducts with the air gap leads to a non-recoverable pressure loss arising from the flow separation and subsequent turbulent mixing. The effect of this combining flow increases the resistance of flow passing through the air gap. The phenomenon of combining flow has been studied by Chong [12] using CFD and experimental methods. The CFD simulations demonstrate that the deflection of rotor ducts flow is strongly dependent on the rotor speed.

Besides rotor speed, the experimental investigation found that the combining flow loss is also affected by the proximity of the rotor ducts to the air gap. When compared to rotor ducts at H/r_i of 0.5, the rotor ducts at H/r_i of 0.75 are much closer to the air gap and hence the effect of combining flow is more significant. Similarly to the air gap and rotor duct entrance loss, the coefficient of combining flow loss can be characterized using the rotation ratio (V_T/U) where V_T is the rotor outer surface velocity and U is the mean axial flow velocity in the air gap. Also, based on the experimental results of Chong [12], the number of rotor ducts has some influence on the combining flow loss.

For the case of 12 rotor ducts at H/r_i of 0.75, the combining flow loss coefficient for the air gap can be correlated as:

$$\begin{aligned} V_T/U > 1, & \quad K_{CF} = 0.31(V_T/U)_{gap} - 0.31 \\ V_T/U \leq 1, & \quad K_{CF} = 0 \end{aligned} \tag{3.55}$$

The combining flow loss increases linearly with the rotation ratio, but it can be neglected when the rotation ratio is less than approximately 1.

For the case of six rotor ducts at H/r_i of 0.75, the combining flow loss coefficient for the air gap can be correlated as:

$$\begin{aligned} V_T/U > 2.5, & \quad K_{CF} = 0.3 \\ V_T/U > 1, & \quad K_{CF} = 0.17(V_T/U)_{gap} - 0.14 \\ V_T/U \leq 1, & \quad K_{CF} = 0 \end{aligned} \tag{3.56}$$

When the rotation ratio is greater than unity, the coefficient of combining flow loss increases with the increase in rotation ratio and it becomes constant after the rotation ratio is greater than 2.5. It means that the combining flow reaches a saturated value and the flow resistance of the flow passing through the air gap is not affected by any further increase in the rotation ratio.

To eliminate the combining flow, a flow guard is proposed by Chong [12] as shown in Figure 3.19. The flow guard is a tube extending from the rotor end surface which is concentric with the rotor. It was placed between the rotor ducts and rotor–stator gap to stop the rotor duct flow from interrupting the air gap flow. The validity of using a flow guard to eliminate the combining flow has been verified through experimental investigation. Since the flow rate passing through the rotor ducts for a given pressure is the same for the case with and without the flow guard, the flow resistance caused by the flow guard is negligible

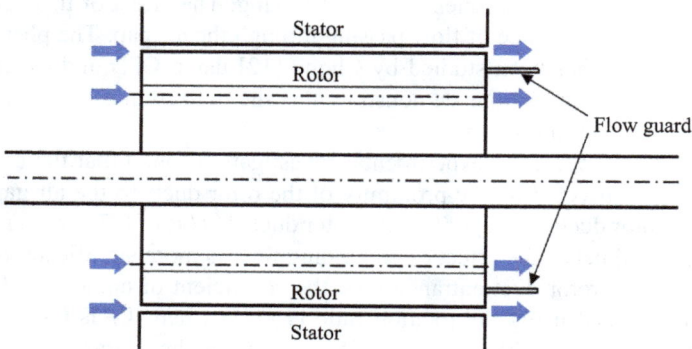

Figure 3.19 Flow guard

3.6　Flow over rotating discs

Axial flux permanent magnet (AFPM) machines are an attractive solution for many applications such as hybrid and electric vehicles and wind turbine generators because of their high torque density and efficiency.

For a self-cooled AFPM machine, the air adjacent to the rotor is dragged by the rotor due to the no-slip condition and driven radially outwards by centrifugal force. Hence, the rotor movement creates a pressure difference to draw the air from the surroundings towards the rotor centre to replace the cooling air which has been pumped out. Moreover, protruding magnets on the rotor assist the pumping effect. The heat generated in the machine is transferred to the radial outflow by forced convection. Two main flow regimes are exhibited in a flat rotating disc facing a stationary disc: Batchelor flow as defined by Batchelor [15] and Stewartson flow as defined by Stewartson [16]. Batchelor flow normally describes the flow structure with the radial outflow adjacent to the rotor, replaced by the radial inward flow adjacent to the stator from the periphery to satisfy the conservation of mass. The distinction between Batchelor flow and Stewartson flow is that Batchelor flow has an inviscid rotating fluid core between separate tangential boundary layers of the rotating and stationary discs, respectively, and at almost zero radial velocity. Batchelor flow is more likely to occur in enclosed systems, whereas Stewartson flow tends to appear in open and throughflow ventilated systems. Convective heat transfer has been investigated for open rotor–stator systems in Refs [17–19]; and enclosed rotor–stator systems in Ref. [20]. The convective heat transfer correlations that have been developed for these systems show that the Nusselt number mainly depends on the rotational Reynolds number and the gap ratio which is defined as the ratio of gap size over the rotor outer radius.

References

[1]　Rumpel G. and Sondershausen H. D. Mechanics. In: Beitz W. and Küttner K. H., editors, *Dubbel Handbook of Mechanical Engineering*, 1st edn. London: Springer; 1994.

[2]　Bhatti M. S. and Shah R. K. Turbulent and transition flow convective heat transfer in ducts. In Kakac S., Shah R. K., and Aung W., editors, *Handbook of Single-Phase Convective Heat Transfer*, 1st edn. New York, NY: Wiley-Interscience; 1987 (Chapter 4).

[3]　Idelchik I. E. *Handbook of Hydraulic Resistance*, 4th edn. Redding: Begell House, Inc.; 2007.

[4]　Miller D. S. *Internal Flow Systems*, 1st edn. UK: BHRA Fluid Engineering; 1978.

[5]　Staton D. A. and Cavagnino A. 'Convection heat transfer and flow calculations suitable for electric machines thermal models'. *IEEE Trans. Ind. Electron.* 2008; 55(10):3509–3516.

[6]　Osborne W. C. and Turner C. G. *Woods Practical Guide to Fan Engineering*, 2nd edn. Colchester, UK: Woods Colchester Ltd.; 1960.

[7]   Lightband D. A. and Bicknell D. A. *The Direct Current Traction Motor: Its Design and Characteristics*. London, UK: Business Books Ltd.; 1970.

[8]   Taylor G. I. 'Stability of a viscous liquid contained between two rotating cylinder'. *Proc. R. Soc., Lond., Ser. A*. 1923; 223:239–343.

[9]   Jung W. M., Kang S. H., Kim W. S., and Choi C. K.. 'Particle morphology of calcium carbonate precipitated by gas–liquid reaction in a Couette–Taylor reactor'. *Chem. Eng. Sci*. 2000; 55(4):733–747.

[10]   Becker K. and Kaye J. 'Measurements of diabatic flow in an annulus with an inner rotating cylinder'. *J. Heat Transf*. 1962; 849(2): 97–104.

[11]   Kaye J. and Elgar E. 'Modes of adiabatic and diabatic fluid flow in an annulus with an inner rotating cylinder'. *Trans. ASME*. 1958; 80: 753–765.

[12]   Chong Y. C. *Thermal Analysis and Air Flow Modelling of Electrical Machines*. Ph.D. Dissertation, School of Mechanical Engineering, Institute for Energy Systems, Edinburgh University, Edinburgh; 2015.

[13]   Yamada Y. 'Resistance of a flow through an annulus with an inner rotating cylinder'. *Bull. Jpn. Soc. Mech. Eng*. 1962; 5(18):302–310.

[14]   Johnson A. R. and Morris W. D. 'An experimental investigation into the effects of rotation on the isothermal flow resistance in circular tubes rotating about a parallel axis'. *Int. J. Heat Fluid Flow* 1992; 13(2):132–140.

[15]   Batchelor G. 'Note on a class of solutions of the Navier-Stokes equations representing steady rotationally-symmetric flow'. *Quart. J. Mech. Appl. Math*. 1951; 4(1):29–41.

[16]   Stewartson K. 'On the flow between two rotating coaxial discs'. *Math. Proc. Camb. Philosph. Soc*. 1953;49(2):333–341.

[17]   Owen J. and Rogers, R. *Flow and Heat Transfer in Rotating-Disc Systems, Vol. 1: Rotor–Stator Systems*, 1st edn. Taunton: Res. Stud. Press; 1989.

[18]   Boutarfa R. and Harmand S. 'Local convective heat transfer for laminar and turbulent flow in a rotor-stator system'. *Exp. Fluids* 2004; 38(2): 209–221.

[19]   Howey D. A., Holmes A. S., and Pullen K. R. 'Measurement and CFD prediction of heat transfer in air-cooled disc-type electrical machines'. *IEEE Trans. Ind. Appl*. 2011; 47(4):1716–1723.

[20]   Daily J. and Nece R. 'Chamber dimension effects on induced flow and frictional resistance of enclosed rotating disks'. *Trans. ASME, J. Basic Eng*. 1960; 82(1):217–232.

Chapter 4

Thermal modelling of electrical machines

Thermal modelling and analysis is an important topic for the electrical machine design process due to the demands for high machine power output with reduced weight, reduced cost and increased efficiency. Also, there is a strong interaction between electromagnetic and thermal design. For instance, the losses are dependent on machine temperatures and vice versa; an increase in magnet temperature will lead to a decrease in flux and thus reduce the output torque; the electrical resistance of copper windings increases with temperature and hence elevated winding temperatures give much higher copper losses and hence reduce the machine efficiency. Therefore, it is essential for electrical machine engineers to consider both electromagnetic and thermal design before an optimum design can be obtained.

The current approaches in the thermal analysis of electrical machines are the analytical lumped-parameter thermal network (LPTN) method, numerical finite-element analysis (FEA), and computational fluid dynamics (CFD). From the literature review of electrical machine thermal analysis, the LPTN is shown to be a well-established method in predicting the thermal performance of electrical machines. LPTN has the advantages of fast computation speed, even for thermal transients, therefore it is an attractive modelling tool for traction motor drive cycle analysis. When compared to numerical methods, sensitivity analysis is more easily implemented using LPTN. This makes it suitable for electrical machine design optimization. In product development, LPTN is useful for all areas of the workflow and not just detailed design. It can also be used by application engineers to do motor type and topology selection for initial sizing. Besides, the LPTN method is commonly used by system engineers for system modelling through reduced order models. Test and design engineers can perform LPTN model calibration based on test data and the calibrated models are further used for design optimization and improvement.

It is appropriate in this introduction to give an overview of the importance of coolant flow rate, coolant heat transfer coefficient and conduction thermal resistance in determining peak temperatures in a machine. The cooling of an electrical machine can be simplified as the heat transfer problem of a fluid flow through a pipe with a constant wall heat flux. The following formula is commonly used to calculate the heat transfer rate for a given mass flow rate due to the temperature rise of the coolant between the inlet and the outlet:

$$q = \dot{m}c\Delta T \tag{4.1}$$

On the other hand, based on Newton's law of cooling, the convective heat transfer between the coolant and the pipe can be expressed as:

$$q = hA\Delta T \tag{4.2}$$

It is important to note that the ΔT in (4.1) and (4.2) are different. In (4.1), the ΔT refers to the coolant temperature rise from its initial temperature to its final temperature leaving the system. However, the ΔT in (4.2) refers to the temperature difference between the solid surface and the surrounding fluid. The heat transfer equations provide useful insight into the relationship between the fluid temperature and wall temperature, which directly relates to the internal temperature of the electrical machine along the heat flow path. In the case with a constant heat flux boundary condition, as occurs due to the losses in an electrical machine, the fluid mass flow rate strongly affects the outlet temperature of the fluid. For a given inlet temperature, a higher mass flow rate leads to lower outlet temperature and therefore this gives a lower fluid temperature for convective heat transfer along the pipe. As the heat transfer coefficient h is likely to be constant the temperature difference between the wall and fluid temperature is typically the same along the pipe, as shown in Figure 4.1. This explains how a higher mass flow rate results in a lower pipe wall temperature and hence lower internal machine temperature.

The temperature difference between the wall and the fluid in a cooling duct is inversely proportional to the convective heat transfer coefficient between the fluid and the wall as illustrated in Figure 4.1. Higher coolant flow rates will give higher heat transfer coefficients. So higher coolant velocities will also reduce the temperature difference between the wall and the fluid. The peak temperatures within an electrical machine occur within the active solid regions of the machine, typically the windings and iron components of the magnetic circuit where there are alternating fluxes. Heat is transferred by conduction from these regions to the walls where convection heat transfer takes place. Thermal resistances due to conduction are generally of the form $\Delta x/kA$, where Δx is the length of the path for heat conduction from the place where the heat is generated to the cooled wall, k is the thermal conductivity along the conduction path and A is the cross-sectional area across which heat is conducted. This is also illustrated in Figure 4.1. For small electrical machines, the conduction path length Δx is small and the conduction thermal resistance is small. This means that the active parts of the machine are not much higher in temperature than the cooled walls. As machines become larger, the conduction path becomes longer giving a larger conduction thermal resistance. This results in higher temperature in the active solid regions such as the windings. The conduction thermal resistances can be reduced by reducing the conduction path length to the cooled walls and in large machines this is achieved by embedding cooling ducts within the solid regions, for example, in the iron stator core where axial or radial cooling ducts may be placed. These cooling ducts inevitably make the machine thermal design more complex but are a necessity to ensure that peak machine temperatures are kept to appropriate levels.

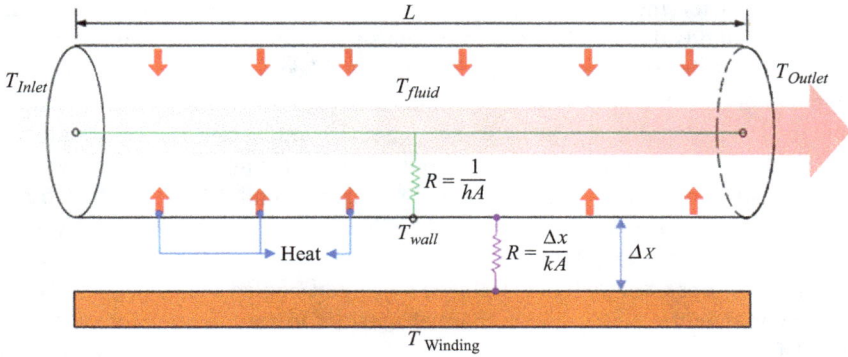

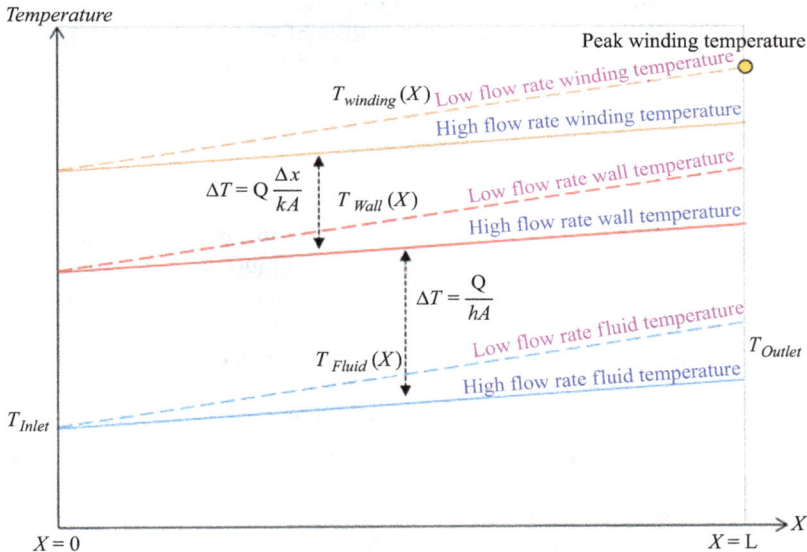

Figure 4.1 *The influence of flow rate on fluid outlet and wall surface temperature under constant heat flux condition, while convective heat transfer coefficient determines the temperature difference between the fluid and pipe wall*

The simple illustrations shown in Figure 4.1 show that the peak temperature in an electrical machine can reduced by three means: minimizing the conduction thermal resistance within the solid regions; maximizing the convective heat transfer coefficient in coolant ducts and; maximizing the coolant flow rates along the ducts.

In this chapter, the authors explain how an equivalent thermal network for an electrical machine is generated. The accuracy of thermal network is dependent upon the appropriateness of the thermal resistance components used.

The heat transfer mechanism in some machine parts is very complicated and empirical methods that can be used to model such components are presented such as for windings, bearings, and the interface gap. As the winding consists of not only copper conductors but also other insulation materials such as enamel, slot liner and impregnation material that fills the gap between the conductors, the FEA method is very useful to predict the winding conduction heat transfer. Convective heat transfer is still the most complex issue and requires an understanding of fluid flow within electrical machines. As described in Chapter 3, rotation significantly influences flow resistance and hence convective heat transfer coefficients in electrical machines. For open ventilated and closed air circuit-cooled machines, a flow network is commonly coupled with a thermal network to form a complete analytical thermal-fluid modelling method to accurately predict electrical machine thermal performance. Alternatively, due to improvements in CFD capabilities and computer power, the CFD method allows the analysis of complex fluid flow to be performed. In addition, the authors have provided case studies of six different cooling methods that are commonly used by electrical machines in different applications.

4.1   Modelling technique – lumped parameter thermal network

The equivalent thermal circuit model has been used in the past for simple motor sizing. The simplified equivalent thermal circuit in Figure 4.2 allows motor engineers to understand how the thermal resistance (R) and the capacitance (C) affect the machine temperature rise from the ambient. Similar to an electrical network, thermal resistance, heat flow and temperature difference are analogous to electrical resistance, current, and voltage.

For the simplified thermal circuit in Figure 4.2, the heat dissipation of the electrical machine is represented by equivalent thermal resistances which can be determined empirically from the machine losses and temperature difference between machine and ambient at steady state. This simplifies the process of detailed calculation of all the heat transfer mechanisms involved (conduction, convection, and radiation)

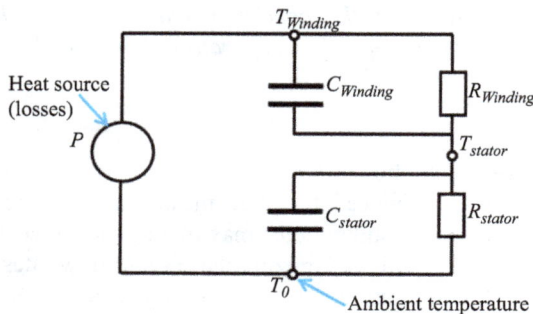

Figure 4.2   Simplified equivalent thermal circuit

which is difficult to predict or measure. The winding temperature can then be predicted for given losses (P) as:

$$T_{winding} = T_0 + RP \tag{4.3}$$

With thermal capacitances, the simplified thermal circuit in Figure 4.2 makes it possible to calculate the general heating and cooling of the machine as a function of time. Then the temperature rise of the winding relative to the ambient can be estimated using the formula given below:

$$T_{winding} - T_0 = (T_s - T_0)\left(1 - e^{-\frac{t}{\tau}}\right) \tag{4.4}$$

$T_s - T_0$ is the winding steady-state temperature rise. τ is the thermal time constant in seconds which is indicative of the rate of response to a change in temperature and is defined as:

$$\tau = RC \tag{4.5}$$

The thermal capacitance (C) is calculated based on the material density (ρ), volume (V), and specific heat capacity (c) as:

$$C = \rho V c \tag{4.6}$$

Figure 4.3 illustrates a typical temperature rise of an electrical machine under load. The time taken by stator winding to reach a steady-state temperature is shorter when compared to the stator yoke because the stator yoke has larger thermal time constant than the stator winding.

The empirically determined equivalent thermal resistance in (4.3) comprises heat transfer phenomena such as conduction heat transfer through composite components (winding, bearings, and steel laminations), interface gaps between machine components, convection cooling around the end windings, convection

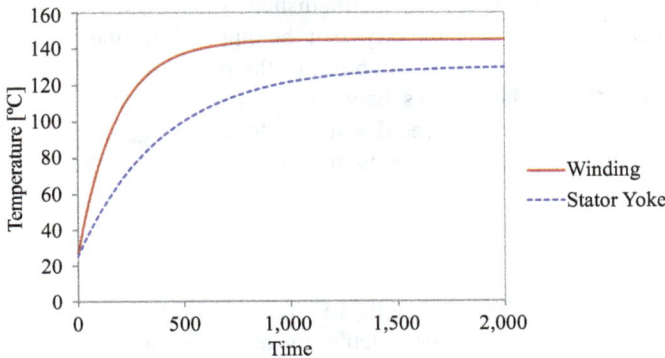

Figure 4.3 *The influence of thermal time constant on winding and stator yoke temperature rise*

cooling from the housing fins to ambient, and radiation from internal and external surfaces of the machine. Consequently, the empirical values of the resistance are limited to certain ranges of machines that have similarity in: materials used, manufacturing process, and geometric design. Any change in those factors will affect the resistance, so the single resistance thermal circuit becomes very inaccurate when trying a new design.

However, besides the winding temperature, there is a requirement to model the temperature of other machine components such as the magnets, bearings, and the winding with higher thermal resolution to identify hot spots. Therefore, a multiple node thermal network is required. Also by having multiple nodes, heat transfer phenomena can be modelled separately by individual thermal resistances. The calculation of individual thermal resistances is presented in later sections. It has been demonstrated that the accuracy of thermal calculation strongly depends on the number of nodes in the equivalent thermal circuit and the accuracy the thermal resistance computation [1]. Moreover, a multiple node model provides insight into where the thermal barriers are that cause the temperature rise and it allows machine designers to concentrate the design effort to reduce the corresponding thermal resistances.

In the multiple node model, a complete electrical machine is represented by an equivalent thermal network which consists of a number of thermal nodes that are connected by thermal resistances. The thermal network lumps together the machine components with similar temperatures into discrete nodes. Power sources and thermal capacitances are distributed between the nodes. The power sources are due to machine losses which can be copper losses, stator and rotor iron losses, magnet loss for permanent magnet machines, rotor cage loss for induction machines, bearing, and windage losses. The losses are placed at the nodes where they are generated. Hence, the thermal resistance circuit describes the main paths for heat flow, enabling the temperatures of the main components within the machine to be predicted for a given loss distribution. Figure 4.4 illustrates a schematic diagram of an electric machine geometry cross-section overlaid with the equivalent thermal network. The thermal resistances are distinguished by the same color code as the machine geometry. Cyclic symmetry can be applied to the thermal network depending on the number of slots and poles of the machine.

In general, all the heat flows between the thermal nodes can be modelled individually by thermal resistances due to conduction (R_{cond}), convection (R_{conv}), and thermal radiation (R_{rad}). For conduction, it is defined as follows:

$$R_{cond} = \frac{L}{kA} \tag{4.7}$$

where L is the heat transfer path length, k is the material thermal conductivity, and A is the cross-sectional area. The calculation of conduction is very complex for certain components, e.g. windings, bearings, interfaces between machine components, etc. Approaches that can be used to model those complexities are provided in the sections below.

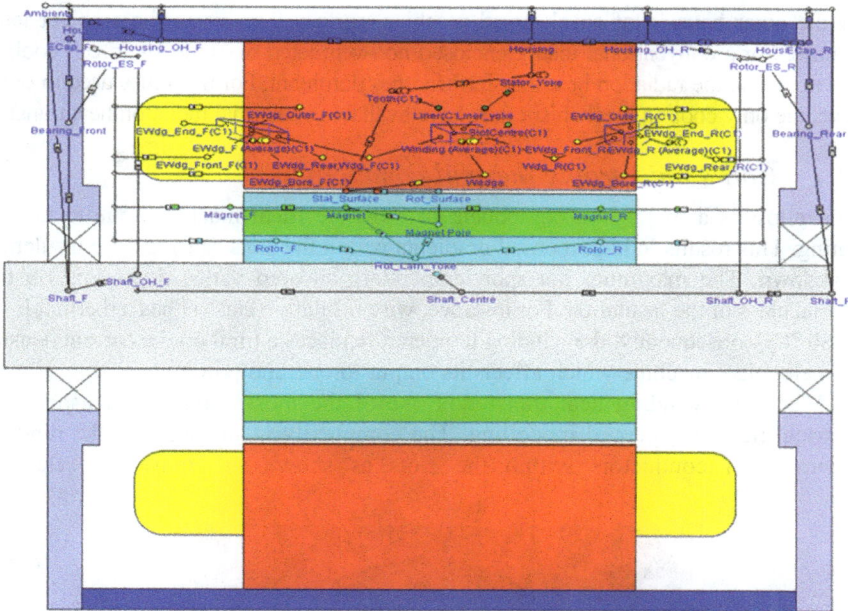

Figure 4.4   Schematic view of machine geometry cross section overlaid with equivalent thermal network using Motor-CAD software

Convective cooling is a common heat transfer mechanism used by electrical machines. For convection, the thermal resistance is defined as follows:

$$R_{conv} = \frac{1}{h_c A} \tag{4.8}$$

where A here is the surface area rather than cross-sectional area as in the conduction thermal resistance. h_c is the convective heat transfer coefficient which is determined empirically from experimental measurement or from CFD simulation. A range of empirical correlations that are applicable to electric machines has been provided in Chapter 2.

For thermal radiation, the thermal resistance is defined as follows:

$$R_{rad} = \frac{1}{h_r A} \tag{4.9}$$

where h_r is the radiation heat transfer coefficient and A here is the area of radiating surface. h_r is calculated using the formula below:

$$h_r = \frac{\sigma \varepsilon F_{1-2}(T_1^4 - T_2^4)}{T_1 - T_2} \tag{4.10}$$

where ε is the emissivity of radiating surface ranging from 0 to 1, σ is the Stefan-Boltzmann constant (5.669×10^{-8} W/m^2/K^4), F_{1-2} is the view factor between surfaces 1 and 2. Surface 1 is the radiating surface while surface 2 is the surface radiated to.

For the outer housing of an electrical machine, surface 2 refers to the ambient temperature. For a conventional totally enclosed non-ventilated machine, as forced cooling does not exist, the radiation heat transfer becomes dominant. For machines used in outer space, the only cooling mechanism they have available is thermal radiation heat transfer.

4.1.1   Coil's anisotropic thermal conductivity

Copper loss is a major loss component of most electrical machines due to Ohmic heating. This results in high thermal gradients within the slots causing wire insulation breakdown. The maximum hot spot temperature allowed varies depending on the thermal class of the insulation. For instance, wire insulation class H has a thermal limit of 180 °C. Consequently, the winding temperature places a limit on the current density of an electrical machine which affects its output torque and power.

Modelling winding heat transfer is one of the most complicated aspects of electrical machine thermal modelling. The complexity is attributed to the random placement of conductors within the slots as shown in Figure 4.5. Besides

Figure 4.5   Cross-section of winding wires

conductors, within the slots, the thermal properties of the other insulation materials (wire insulation, phase divider, impregnation, slot liner, etc.) vary. Typically, the thermal conductivity of insulation materials is in the range of 0.1–1 W/m/K, while the pure copper thermal is 400 W/m/K. Furthermore, air can be trapped in the space between conductors and the amount of air void depends on the winding impregnation process. Air thermal conductivity is 0.026 W/m/K at ambient temperature which is much lower than the insulation. Therefore, the vacuum pressure impregnation (VPI) process is recommended to eliminate air voids for better winding heat dissipation. In electrical machine impregnation process, the quality of impregnation is indicated by impregnation goodness factor from 0 to 1. The value of 1 indicates the conductor space is filled 100% by impregnation material without any air voids.

In analytical thermal modelling, the winding can be represented as a bulk volume with anisotropic thermal conductivity. The conduction heat transfer in radial, tangential, or axial direction can be modelled using an equivalent thermal conductivity in each direction based upon the ratio of materials (conductor, wire insulation, impregnation, and air void) and their thermal conductivities. The appropriateness of using equivalent thermal conductivity to model the winding heat transfer has been studied experimentally in Ref. [2] and numerically in Ref. [3]. In the axial direction, based upon the parallel arrangement of winding materials, the equivalent thermal conductivity along the winding is mainly dominated by the conductor material (often copper). In the radial and tangential directions, the equivalent thermal conductivity across a coil is much more difficult to estimate than for the axial direction. However, it can be approximated using the empirical formulations proposed by Haskin and Shtrickman [4] and Milton [5] which is as a function of wire slot fill factor as shown in Figure 4.6. The FEA method is very useful to model the winding heat transfer and a comparison between empirical methods and the FEA method is also given in Figure 4.6. The slot fill factor is also affected by winding

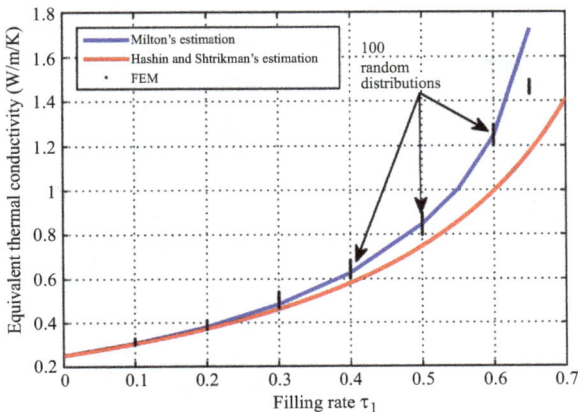

Figure 4.6   The variation of the wire slot fill factor with equivalent thermal conductivity (radial and tangential direction) [3]

type, e.g. distributed winding, concentrated winding, form wound winding, hairpin winding, and Litz wire.

4.1.2   Winding heat transfer

Most of the thermal resistive network is developed from a one-dimensional (1D) solution concept. However, heat transfer in the winding is a more three-dimensional (3D) problem as the winding losses can dissipate to stator yoke radially, to stator tooth circumferentially, and to end windings axially. To model the winding, the cuboidal element and arc-segment element proposed by Refs [6,7] as shown in Figure 4.7 can be applied to the thermal network so that the direction heat flow takes place does not have to be pre-determined beforehand as it depends where the cooling is.

In the two resistor network, R_1 and R_2 are based on the steady-state heat diffusion equation for zero internal heat generation and it will give accurate results in this case. However, where there is internal heat generation within a conducting solid, an additional resistance term R_3 is required. This is shown for the cuboidal and arc-segment elements shown in Figure 4.7 where an additional resistor R_3 is introduced for each direction to form so-called T-networks. These T-networks are required to properly model internal heat generation to give an accurate value for the mean temperature within an element. In the T-network, the additional resistance R_3 has a *negative* value that is one third of the value of the resistances R_1 and R_2. The reason for this negative resistance is to allow for the internal heat generation not occurring at the center of an element, but being uniformly distributed throughout.

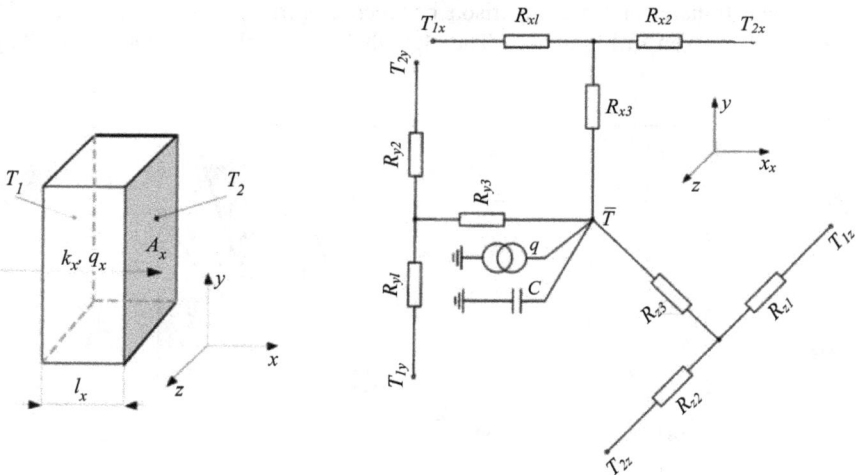

Figure 4.7   *3D thermal modelling of electrical machines. 3D network representation of the general cuboidal element [6]. 3D network representation of the general arc-segment element [7]*

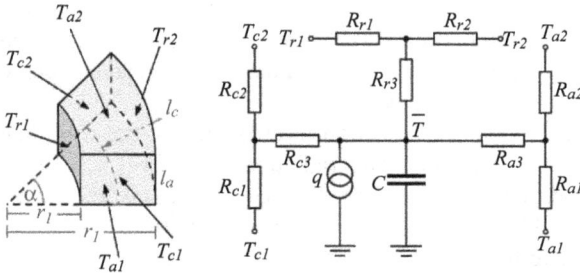

Figure 4.7 (*Continued*)

Heat that is generated away from the center of the element has a shorter distance to travel to reach the sides of the element and so has a lower effective thermal resistance in traveling out of the element. The thermal resistance R_3 thus has a negative value prevent the mean temperature in the element being over predicted. These T-networks must be used in modelling conduction within parts of a machine, such as the windings and stator teeth where there is a significant internal heat generation [6,7].

In addition, the anisotropic thermal conductivity of the coil as presented in Section 4.1.1 can be applied within the cuboidal element and arc-segment element for accurate thermal calculations. This 3D thermal modelling technique has been applied in Motor-CAD software for complex components like the winding of both active winding in the slots and end windings as shown in Figure 4.7.

4.1.3 End-space cooling

Convective cooling within the endcaps of an electric machine is quite complex due to the difficulty in accurately predicting the fluid flow over, and the heat transfer coefficient from, the different surfaces within the end space, i.e., end winding, endcaps, housing, and rotor. The fluid flow depends on many factors including the shape, length, and winding type of the end winding, the surface finish of end-space surfaces and the fanning effect of the rotor. The fanning effect can be increased by fitting a fan or attaching wafters or axial blades to the ends of the rotor. Wafters or blades are often incorporated on the end ring of an induction motor squirrel cage.

The heat transfer phenomena in the end cap region of totally enclosed air-cooled motors has been analyzed by several authors [8,9,45–47], and they have converged on a general formulation for the computation of the heat transfer coefficient:

$$h = k_1[1 + k_2 V^{k_3}]$$ (4.11)

where V is a local air velocity for the particular surface in contact with the end-space air (for instance the peripheral speed of the wafters could be used multiplied by some efficiency factor) and k_1, k_2, and k_3 are empirical curve fitting constants. The empirical formula is only applicable for end-space surfaces cooled by air at atmospheric pressure (at sea level). The influence of altitude on end-space cooling can be considered through a pressure correction term which is given in Chapter 2.

Table 4.1 and Figure 4.8 show six published empirical correlations for a range of totally enclosed fan-cooled motors.

The end-space cooling correlation models include the combination of natural convection, forced convection, and radiation. The term k_1 accounts for natural convection and radiation when the reference velocity is zero. The term $k_1 k_2 V^{k_3}$ accounts for the added forced convection due to rotation.

It is possible to build a special motor prototype to help measure the end-space cooling heat transfer. For example, the induction motor rotor shown in Figure 4.9 has its rotor laminations and active part of the cage replaced by a plastic cylinder. This removed the rotor magnetic and electrical losses when the rotor rotates. The end-rings and wafters remain as they influence the end-space airflow. The external fan is also removed if present.

The prototype motor has had extensive thermal sensors added to measure the end-winding temperature at different positions. The motor prototype is driven

Table 4.1   Typical TEFC end-space cooling
             correlation data

	k_1	k_2	k_3
Mellor	15.5	0.39	1
Schubert [EW]	15	0.4	0.9
Schubert [Bg]	20	0.425	0.7
Stokum	33.2	0.0445	1
Di Gerlando	40	0.1	1
Hamdi	10	0.3	1

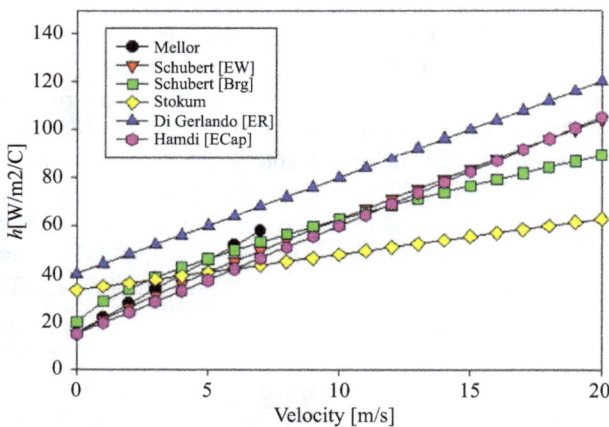

Figure 4.8   Published end-space convection heat transfer correlations [10]

externally and losses applied to the stator winding using a dc supply. The only loss within the machine is the stator copper loss with a small amount of bearing loss. The measured temperature data for different rotational speeds and imposed stator copper loss can then be used to calibrate k_1, k_2, and k_3. This test has been used to measure typical k values for three different totally enclosed fan-cooled induction motors with the results reported in Table 4.2 and Figure 4.10.

The dashed lines shown in Figure 4.10 are the upper and lower limits of the h value reported in Figure 4.8. The results for the three new motors have similar trends to the results in Figure 4.8. In general, the newer motor designs as measured

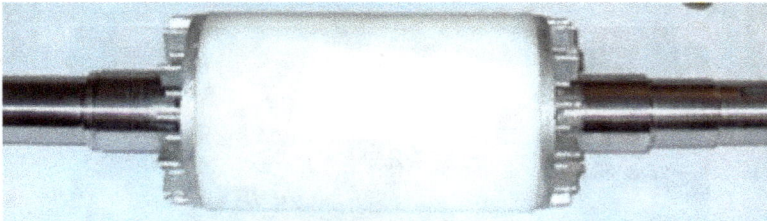

Figure 4.9   Plastic rotor used for end-space cooling experimentation

Table 4.2   End-winding heat transfer coefficients

	k_1	k_2	k_3
Motor 1	38	0.7	1
Motor 2	20.5	0.12	1
Motor 3	35.3	0.1	1

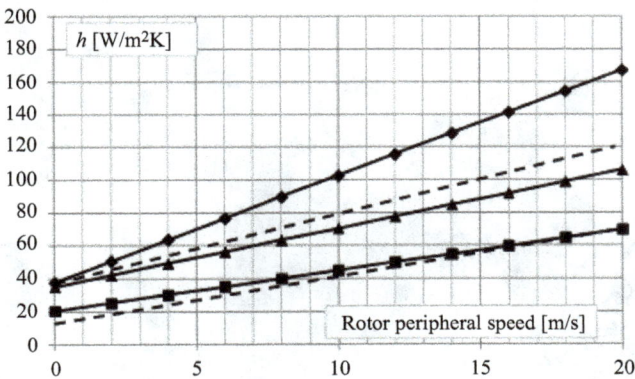

Figure 4.10   Measured end-winding heat transfer coefficient for three totally enclosed induction motors

in the tests shown in Figures 4.8 and 4.10 show an increase in end-winding heat transfer coefficient compared to the older designs. In Figure 4.8, the curves by Schubert and Stokum were published in the sixties, while those of Di Gerlando, Mellor, and Hamdi were published in the nineties.

Tests have also been carried out on an open drip proof "ODP" motor with a plastic rotor. Tests were performed with the frame and end-cap opening open (ODP) and blocked (TE) as shown in Figure 4.11. Comparing the test results for the open and closed configurations, it is possible to quantify the ventilation effect due to openings in the frame and end-caps. In addition, removing the end ring from the

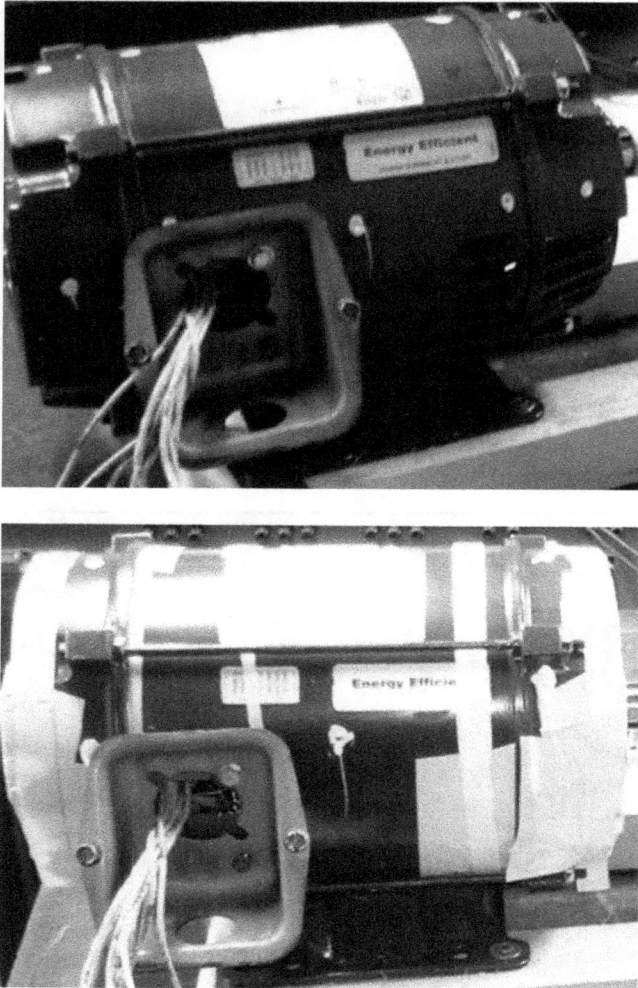

Figure 4.11   Motor in ODP and TE configuration

plastic rotor allows the ventilation effects due to the wafters to be quantified. Figure 4.12 and Table 4.3 compare the results. As expected, the frame and end-cap openings give a significant increase in end-space cooling. The rotor wafters also increase the end-space cooling significantly. Vents in the endcap or housing also have the benefit of allowing cooler air into the interior of the machine. In the extreme case for a fully open machine, the internal air temperature will be the same as the external air temperature. For a partially open machine, an estimation of the cooling of the end-space air can be made by estimating the air flow rate entering the machine.

When using the end-space heat transfer correlation, it is important to make an estimate of velocity of the fluid on each surface. The inside of the end-winding will have a larger fluid velocity than the outside of the end-winding as it is closer to the rotor. In Motor-CAD software, an empirical approach is used for the estimation of the fluid velocity on each surface, with past experience factors built into the software. CFD can also be useful to help calibrate a lumped circuit model for the end-space cooling as demonstrated in [11] and presented in Section 5.2.4.2.

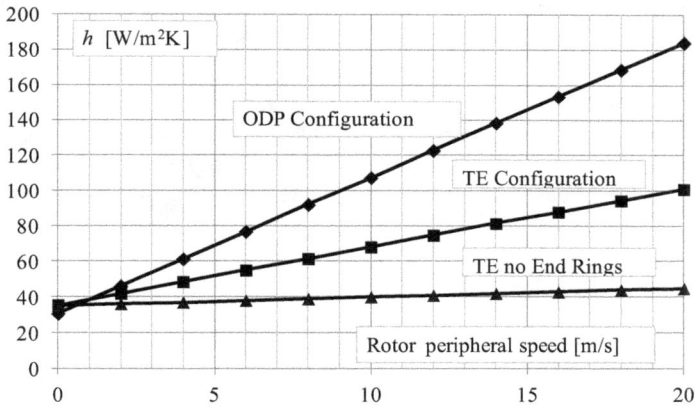

Figure 4.12   Heat transfer coefficient in ODP, TE, and TE no end rings configurations

Table 4.3   Data for ODP, TE, and TE with no end rings configurations

	k_1	k_2	k_3
ODP configuration	30.7	0.25	1
TE configuration	35.3	0.093	1
TE no end rings	35.0	0.014	1

4.1.4   *Bearing losses and heat transfer*

Accurate modelling of bearings becomes more important in high-speed applications and in the more compact design of electrical machines. High-speed operating conditions can lead to significant frictional loss in the bearings. The bearing loss is equal to the bearing frictional torque multiplied by the rotational speed. Bearing loss is also influenced by the bearing type, lubrication method, type and quantity of lubricant, bearing sealing arrangement, and the loads applied to the bearing. Bearing loss models are complex and more information on the methods that can be employed to estimate the bearing loss are: models specified by International standards [12], German standards [13], and in particular models provided by bearing manufacturers (e.g. SKF [14]).

The heat generated q can be transmitted from the bearing as follows:

$$q = q_s + q_L \tag{4.12}$$

q_s is the heat flow from bearing seat area by conduction based on the temperature difference between housing and bearing. It is calculated as follows:

$$q_s = k_q A_s \Delta T \tag{4.13}$$

The thermal conductance (k_q) depends on the way the bearing is mounted and is also a function of the housing design, size, and material, it may differ widely and normally in the range between 200 W/m^2 K and 1,000 W/m^2 K for bearing seat area (A_S) up to 50,000 mm^2.

q_L is the heat convected from a bearing where there is a through flow of lubricant. V_L is the volumetric flow rate, ρ_L is the density and c_L is the specific heat capacity of the lubricant. It depends on the temperature difference between the

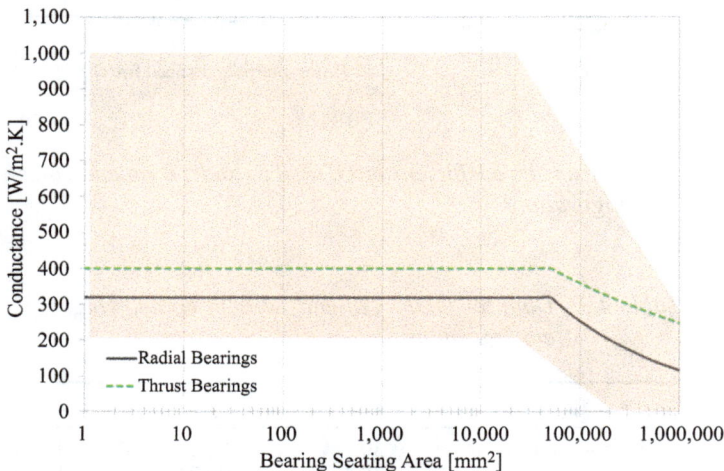

Figure 4.13   Thermal conductance as a function of bearing seating area

lubricant supply temperature and temperature of lubricant discharge, which is usually at the bearing temperature:

$$q_L = c_L \rho_L V_L \Delta T \tag{4.14}$$

A very simple thermal model is proposed in [10] using a simple bearing equivalent thermal resistance based on static tests on motors using similar bearings to those to be used in the final design. Taking into account the mechanical structure of a bearing, a reasonable bearing thermal model can be obtained considering the bearing equivalent to an interface gap. Reasonable values of the equivalent gap can be obtained from thermal tests on specially prepared motors. The motor under test requires special end shields with holes for measuring the temperature of the inner and outer bearing races as shown in Figure 4.14.

The thermal test procedure starts from an accurate calibration of a motor thermal model when the machine is supplied with DC supply. The calibration has to be done taking into account all the critical thermal parameters involved in the conduction and in the natural convection thermal exchange such as housing-lamination interface gap, winding-insulation system, housing natural convention, and radiation thermal resistances. Obviously, the front and rear bearing have to be included in the model as an equivalent circular interface gap having the bearing geometrical dimensions (inner and outer diameter and bearing thickness).

After the thermal model calibration, a locked rotor test using a three phase sinusoidal supply is performed. Under locked rotor conditions, there are no iron or mechanical losses and only stator and rotor joule losses are present. The rotor losses can be computed as the difference between the stator input power minus the stator joule losses. Starting from the DC calibrated motor thermal model, it is possible to modify the front and rear equivalent bearing interface gaps until the temperatures across bearing thermal resistances match those measured in AC locked rotor thermal test.

The obtained equivalent bearing interface gap values for three induction motors are reported in Table 4.4.

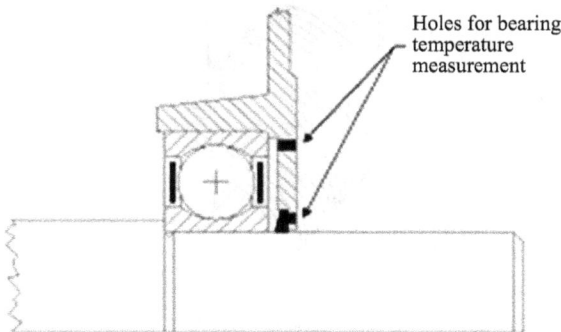

Figure 4.14   End shields for the bearing race temperature measurement

Even if the results reported in Table 4.4 have no general validity, an equivalent interface gap around 0.3 mm can be used as a starting value in a first approach thermal analysis.

4.1.5   Lamination stack heat transfer

The stator and rotor magnetic circuits are most often laminated to reduce induced losses due to circulating currents. However, a significant amount of electromagnetic loss can still be generated in the individual steel laminations – iron loss. Therefore, it is necessary to understand the orthotropic thermal conductivity of lamination stack for accurate thermal modelling of electrical machines. Basically, the effective thermal properties of the lamination stack need to be determined for both in-plane (shown in Figure 4.15 as the x–y plane) and through-stack (shown in Figure 4.15 as the z-direction).

For the through-stack (axial direction), the effective thermal conductivity (k_{eff}) of lamination stack can be determined by assuming the lamination material thermal resistance (R_{lam}) and inter lamination contact resistance ($R_{interlam}$) connected in series as:

$$R_{th} = R_{lam} + R_{interlam} \tag{4.15}$$

$$k_{eff} = \frac{\text{stack length}}{R_{th} \times \text{stack area}} \tag{4.16}$$

Table 4.4   Bearing equivalent interface gap

Motor	Inner diameter (mm)	Outer diameter (mm)	Width (mm)	Equivalent interface gap [mm]
4 kW	30	72	19	0.35
7.5 kW	40	80	18	0.23
15 kW	45	100	25	0.40

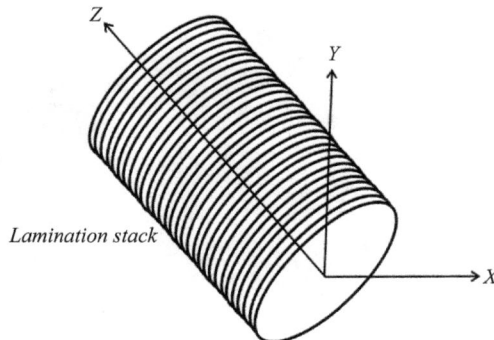

Figure 4.15   Orthotropic thermal property orientation for a lamination stack. Through-stack conductivity is along the z-axis, and in-plane thermal conductivity is along the x–y plane

Based on (4.15), the effective through-stack thermal conductivity is dominated by the inter lamination contact resistance which is affected by inter lamination material, surface finish, steel lamination coating properties, and clamping pressure [15]. To simplify the thermal calculation, all these factors are ultimately reflected by a lamination stacking factor. Also, the lamination thickness has a strong influence on the lamination stacking factor. Thinner laminations lead to more lamination-to-lamination contacts per unit length. This results in a lower stacking factor and thus gives lower effective through-stack thermal conductivity as shown in Figure 4.16. Typically the stacking factor ranges from 0.94 to 0.98. Hence, the lamination stacking factor together with the lamination stack length can be applied into (4.15) for through-stack thermal resistance determination.

For the in-plane direction, the effective thermal conductivity (k_{eff}) can be determined by assuming the lamination material thermal resistance (R_{lam}) and inter lamination contact resistance ($R_{interlam}$) connected in parallel and x is the lamination stacking factor as follows:

$$\frac{1}{R_{th}} = \frac{1}{R_{lam}} + \frac{1}{R_{interlam}} \tag{4.17}$$

$$k_{eff} = xk_{lam} + (1 - x)k_{interlam} \tag{4.18}$$

Due to high lamination stacking factor, the effective in-plane thermal conductivity of lamination stack is dominated by and very close to the lamination material thermal conductivity.

4.1.6 Interfaces

When two solid surfaces are in contact, an interface is formed between two surfaces as explained in Chapter 2. For electrical machines, the interfaces between machine components can potentially cause significant temperature drops. Figure 4.17 shows

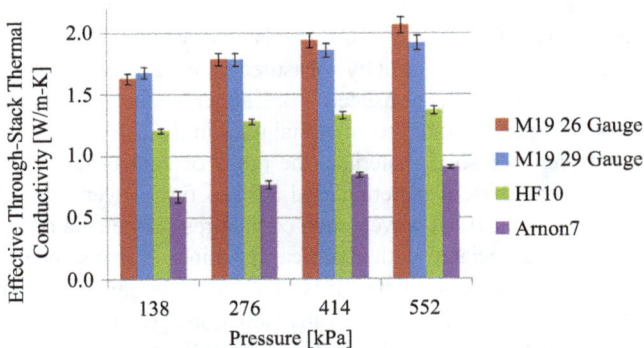

Figure 4.16 *The impact of clamping pressure and lamination material on the effective through-stack thermal conductivity (M19 lamination material in both 26 gauge (0.470 mm) and 29 gauge (0.356 mm), HF10 (0.254 mm), and Arnon 7 (0.178 mm)) [15]*

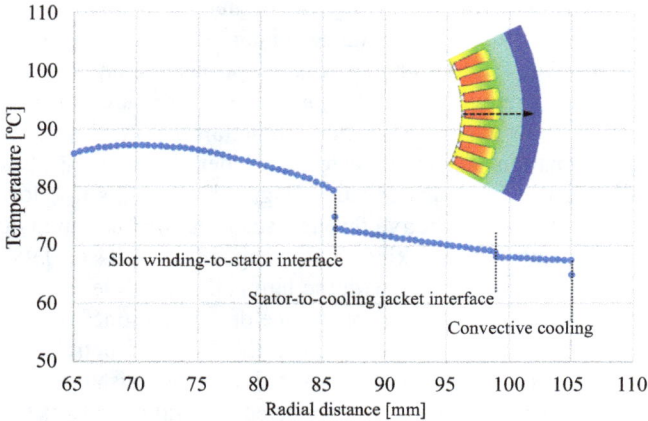

Figure 4.17   Typical impact of interfaces on a housing water jacket-cooled machine

a typical radial temperature profile from the stator core to the housing of a water jacket-cooled machine. This demonstrates the impact of interfaces in a housing water jacket-cooled machine. The heat generated in the conductors has to flow through wire insulation, impregnation, slot liner, coil-to-stator lamination interface, stator tooth and back iron, interface between stator and housing, housing before it dissipates to the cooling channels in the housing. Therefore, care must be taken to determine the thermal contact resistances due to the interfaces along the heat flow path. This is true especially for those machines cooled from the housing, e.g. TENV, TEFC, housing water jacket, etc.

By reviewing the existing literature, the interface can be quantified in three different ways – contact resistance in W/m^2K, contact conductance in m^2 K/W, and equivalent interface gap. The contact resistance is simply the reciprocal of the contact conductance and their values are usually measured using experimental methods. The equivalent interface gap can be translated from either contact resistance or contact conductance depending on the interface material used in the calculation (normally air). For interfaces between two solid materials, the range of contact resistance and interfacial conductance for different materials and surface finish over a range of contact pressure can be found in [10]. However, such contact resistance data does not account for some complexities associated with electrical machines. For instance, the interface gap between lamination stack and housing is likely to be larger than usual due to the laminated stator surface. Several empirical investigations [16–18] have been done on this topic and they have revealed the factors that influence the contact resistance between the housing and stator and also the range of contact resistance values.

1.   Shrink fit pressure.
2.   Housing material. Aluminum is a soft material compared to cast iron and stainless steel which leads to a smaller contact resistance.

3. Stator lamination material.
4. Surface roughness.
5. Preparation of laminated surface and housing surface whether they are machined to be a smooth surface before they are fitted together.
6. The presence of interfacial material such as epoxy, thermal paste or grease which has higher thermal conductivity than air for a bare material joint.
7. Difference in thermal expansion rates between the housing and stator lamination.
8. Machining tolerance.

Typical values of the interface gap between stator lamination and external frame are shown in Table 4.5. In this case, a DC calibration test as proposed in Section 6.4 has been performed on a wide range of four pole TEFC induction motors and the resulting equivalent interface gap between stator lamination and external frame deduced from measurements [19].

The smaller 1.5–15 kW motors have aluminum frames while the larger motors have cast iron frames. There is a significant spread in the data and it is not possible to identify a correlation for the interface gap based on frame material and motor size. The interface gap is strongly influenced by the stator core assembly and by the lamination-frame insertion process. Several of the motor sizes have had multiple tests on different motor batches showing some variation from one motor to another. Even though it is not possible to identify an easy approach to determine the interface gap based just on frame material type and frame size at design time, the calibration data is useful. It can allow a design to be undertaken where the temperature rise can be determined with the assumption of a good fit or bad fit between the stator lamination and frame. A good fit typically being in the region of 0.01 mm and a bad fit approaching of 0.1 mm. A good fit may require an increased investment in a good stator lamination to frame insertion process.

Table 4.5   Equivalent interface gap between
lamination and external frame for four
poles TEFC induction motors

Rated motor power (kW)	Interface gap (mm)
1.5	0.027
2.2	0.049
3	0.03
4	0.04
7.5	0.08
15	0.07
30	0.01
55	0.045
90	0.047
110	0.05
250	0.065
1,000	0.075

4.1.7   Machine losses

In order to predict the temperature distribution in electrical machines, it is important to accurately predict both the magnitude and distribution of the power losses in the machine. Depending on the motor type, several of the following loss mechanisms may be present.

4.1.7.1   Ohmic losses

Ohmic losses or I^2R losses are due to current flowing through the conductors of the motor. These losses are equal to the square of the current multiplied by the resistance of the path through which the current flows. They can be in the stator winding or rotor winding or in the squirrel cage of an induction motor. Cage losses can be minimized by using a copper cage rather than aluminum.

An increase in winding temperature gives an increase in copper resistivity according to the formula:

$$\rho = \rho_{20}[1 + \alpha(T - 20)] \tag{4.19}$$

where $\rho_{20} = 1.724 \times 10^{-8}$ Ωm and $\alpha = 0.00393/°C$

Thus, a 50 °C temperature rise gives 20% increase in resistance and a 140 °C rise gives 55% increase in resistance and ohmic losses. In a permanent magnet motor, an increase in magnet temperature results in a reduction in magnet flux which requires an increase in current to produce the required torque. This will lead to a further increase in ohmic losses.

At high speeds, there may also be an additional component of ohmic losses due to skin and proximity effects. The skin component of loss can be minimized by the use of multiple strands or Litz wire. Proximity losses are more difficult to minimize. Proximity losses are often found in the conductors near to the slot opening and can lead to a severe hotspot temperature at that point. Figure 4.18 shows an example

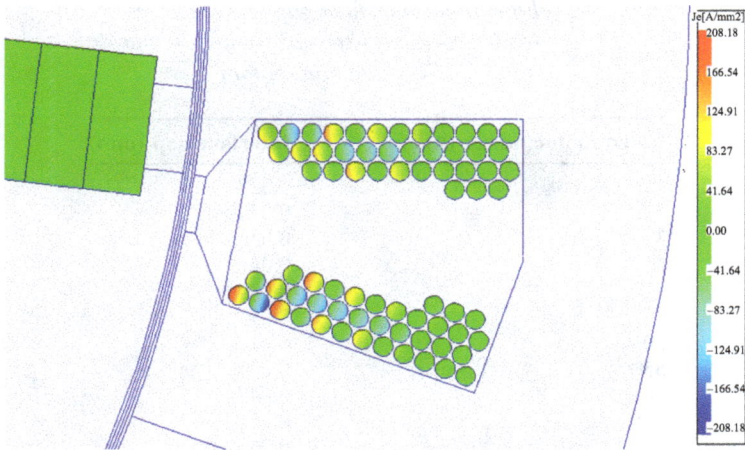

Figure 4.18   Current density in a high speed BPM with spoke magnets

of a BPM motor that has a large component of proximity loss in the conductors at the slot opening at high speeds. This motor has a stator winding having 12 turns per coil and 3 parallel strands in hand. The round conductors have a copper diameter of 1.7 mm. Figure 4.18 shows the instantaneous current distribution obtained using 2D FEA at 1,600 Hz. The maximum current density is 208 A/mm^2 in this case. The ratio of AC to DC resistance is 6 with the DC component of ohmic loss equal to 1,100 W and the AC component of ohmic loss is equal to 6,300 W. This illustrates the possibility of a considerable localized proximity loss even with the use of a stranded winding.

Methods used to reduce proximity loss include twisted wires, use of flat rectangular wire, aluminum wire, and reduced slot fill factor designs with the copper wires pushed toward the bottom of the slot away from the slot opening region.

4.1.7.2 Iron losses

Iron losses are induced in the electrical steel used for the magnetic paths of the motor. They are divided into hysteresis and eddy current losses. Hysteresis losses are due to the changing polarity of the flux in the steel core. Eddy current losses are due to circulating currents induced in the steel core by the changing polarity of the flux. Steel laminations are used to minimize the eddy currents.

The iron loss can be calculated using the generic formula:

$$w_{Fe} = w_{hys} + w_{eddy} = k_h(f, B) \cdot f B^2 + k_e(f, B) \cdot (f B)^2 \tag{4.20}$$

where f is the frequency and B is the local flux density in the electrical steel. k_h and k_e are loss coefficients. The loss coefficients are found by curve fitting the electrical steel manufacturers loss data (loss/kg) as a function of frequency and flux density.

More specific iron loss equations often use the well-known Steinmetz relationship:

$$w_{Fe} = k_h f \cdot B^{(\alpha + \beta \cdot B)} + 2\pi^2 k_e f^2 B^2 \tag{4.21}$$

An alternative referred to as Bertotti's iron loss method has separate terms for hysteresis, classical eddy current and excess loss:

$$w_{Fe} = k_h f B^\alpha + k_e f^2 B^2 + k_{exc} f^{\frac{3}{2}} B^{\frac{3}{2}} \tag{4.22}$$

The steel manufacturers loss data is often measured using perfect laminations on an Epstein frame with perfect sinusoidal current waveforms. In practice, the iron loss in the machine will be larger than that measured on the Epstein frame due to defects such as burs and welds, stress in lamination, and non-sinusoidal waveforms. An iron loss multiplier adjustment is usually used to account for these effects and found from previous experience using the particular electrical steel in similar designs. Iron loss adjustments factors can be as high as 2 or 3.

The equations can be applied using average values for flux density calculated in the stator tooth and back iron. Better results can be obtained using the electromagnetic FEA method with the equations applied to each element.

Motors driven from power electronic converters often have a non-sinusoidal (PWM) voltage waveform that causes increased losses in the lamination steel.

Figure 4.19(a) shows an example of the iron losses calculated in a spoke magnet BPM motor under load. Most of the loss is induced in the stator lamination and can be dissipated via conduction heat transfer to a stator water jacket. There is also a component of iron loss in the laminated rotor pole laminations in this case and needs to be accounted for when designing the cooling system for the rotor in order to minimize the temperature rise of the magnets.

4.1.7.3   Magnet losses

Magnet losses are due to eddy-currents being induced in the magnets. The eddy currents are induced due to dips in the air-gap flux density due to slotting and by current time and space harmonics. Ferrite magnets have a high value of electrical resistivity and so are immune to such losses. NdFeB and SmCo rare-earth magnets have much lower electrical resistivity values and can have a large component of magnet loss. Table 4.6 compares electrical resistivity values for typical materials used in the manufacture of electrical machines.

Figure 4.19(b) shows the average magnet loss over a complete mechanical cycle in a spoke magnet design. In this case, the magnet losses are concentrated at the top of the magnet near the airgap. They must be accounted for when designing the motor cooling system. Magnet losses can be minimized by segmenting the magnet in the radial or axial direction. Axial segmentation is commonly used.

4.1.7.4   Rotor can/sleeve losses

Retaining sleeves are often employed to stop the magnets flying off the rotor due to centrifugal forces in surface magnet BPM motors. If the material of the retainer is electrically non-conducting, e.g. fiber glass or with a very high electrical resistivity, e.g. composite carbon fiber, there are practically no extra losses on the rotor that could affect the magnet temperature. However, non-metallic sleeves have a limited temperature operation to 180 °C. For cost reduction reasons, there are solutions for retaining the magnets using non-magnetic metallic sleeves, e.g. titanium alloys, stainless steel, brass or aluminum.

4.1.7.5   Friction losses

Friction losses can be due to the bearings drag and brush/commutator surface friction in dc motors. Friction losses are proportional to the rotor speed. Bearing losses refer to losses due to friction between the bearing balls and the inner and outer races. Generally, they can be estimated using correlations from manufacturers depending on the geometry of bearing balls, the applied loads and the thermo-physical properties of the lubricant.

4.1.7.6   Stray losses

Stray losses are generally categorized to any additional loss that is not included in one of the above loss types.

4.1.7.7   Windage loss

Many electrical machines currently being designed are high speed and high temperature. Some machines are even operating in fluids different from air and it is

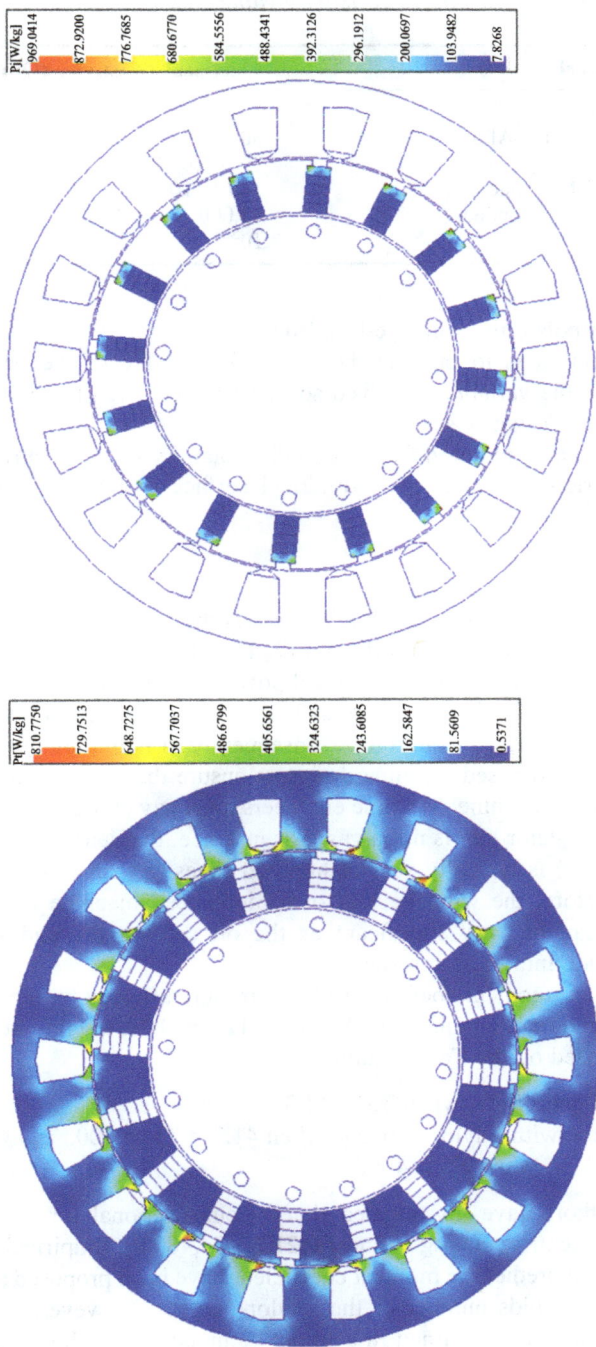

Figure 4.19 (a) Iron losses distribution in a BPM under load condition and (b) distribution of magnet losses in a BPM under load condition

Table 4.6   Electrical resistivity values (Ωm)

Material	Electrical resistivity (Ωm)
Iron	10×10^{-8}
Sm-Co 1-5 Alloys	50×10^{-8}
Sm-Co 2-17 Alloys	90×10^{-8}
NdFeB – sintered	160×10^{-8}
NdFeB – bonded	$14{,}000 \times 10^{-8}$
Ferrite	10^5

difficult to extrapolate the windage loss from commercial machine data. The purpose of this section is to provide theoretical knowledge of the dependence of windage loss on the variables involved so that engineers can predict the windage loss in a machine design.

The power that is required from a cylinder rotating in a concentric cylinder to overcome air friction drag at the cylindrical surface of the rotating cylinder is given by:

$$W = C_m \rho \pi L r^4 \omega_m^3 \tag{4.23}$$

ρ is the fluid density, L is the length of rotor, r is the rotor outer radius, and ω_m is the rotor angular velocity. The coefficient C_m is called the moment coefficient. In rotating electrical machine, this mechanical power is known as windage loss and is due to work done by fluid viscous stress. This power is converted into heat which increases the temperature of the rotor. Consequently, it is important to predict the windage loss of a proposed machine design to ensure that it does not cause detrimental effect to the machine. Machine engineers must pay particular attention if the fluid in the rotor–stator gap is not a gas. For instance, the density of hydraulic oil (800–1,000 kg/m^3) is typically several hundred times to that of air (1.2 kg/m^3 at ambient). Therefore, the windage loss of a high-speed machine that utilizes oil spray cooling can be very significant as the oil cooling the end winding can potentially intrude into the rotor–stator gap.

Moreover, the moment coefficient C_m is strongly affected by the flow regime between stator and rotor annular gap. As shown Figure 4.20, the flow regime can be distinguished based on the Taylor number:

- Laminar Couette flow, when Ta < 41.3.
- Laminar flow with Taylor vortices, when 41.3 < Ta < 400.
- Turbulent flow, when $Ta > 400$.

Several authors have investigated the viscous frictional torque in the flow between two concentric rotating cylinders [21,22–24,48,49]. Empirical correlations that can be used to predict the moment coefficient have been proposed as a function of the air gap Reynolds number or the Taylor number. However, most of these empirical correlations are defined in a smooth annular gap without superimposed axial flow. However, some machines are designed with salient poles and their

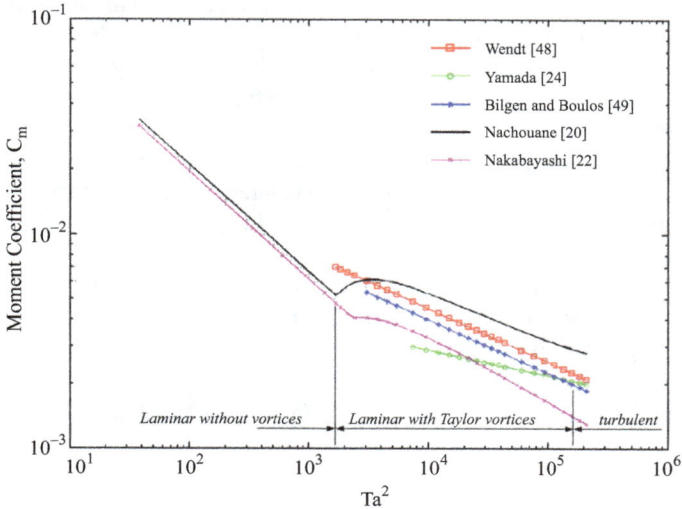

Figure 4.20 *The variation of moment coefficient (C_m) with the square of Taylor number (Ta) [20]*

influence on moment coefficient has been studied in Ref. [23]. The salient pole coefficient is dependent of rotor pole depth to rotor outer radius ratio. For open ventilated machines, the rotational flow in rotor–stator gap is superimposed with axial flow and the investigation in Ref. [24] has shown the impact of axial flow on the moment coefficient.

A comprehensive parametric study of windage loss inside rotor–stator gap has been performed using CFD method [20]. Besides the rotor speed, CFD simulations show that rotor size, air gap size, and even the roughness height have considerable impact on the variation of moment coefficient. A dimensionless analysis is performed based on the parametric study and a set of correlations of the moment coefficient as a function of the studied parameters is proposed in Ref. [20]. Furthermore, the cylindrical surface in the actual rotor–stator gap is not smooth because of stator/rotor tooth. The viscous frictional torque for rotating rough cylinders can be found in Ref. [22].

4.2 TENV cooling

In a totally enclosed non-ventilated (TENV) motor, the cooling is via natural convection and radiation from the motor housing. Servo motors typically use such cooling as they often operate with full torque down to zero speed so a shaft mounted fan would give no cooling at this point. Often the housing will have a smooth outer surface that maximizes the natural convection airflow (like the first two housing types shown in Figure 4.21). Relatively simple convection correlations in Chapter 2 such as those for horizontal cylinders and vertical and horizontal flat plates can be used to calculate the convection cooling from such smooth housings.

If it can be guaranteed that the motor will have a horizontal shaft orientation during operation, then the addition of radial fins to the housing can be used to increase the convective cooling. In such cases, the fin spacing can be optimized for natural convection (third housing type shown in Figure 4.21). More complex convection correlations have been developed to calculate the natural convection in the fin channels.

An example of a sensitivity analysis on the fin spacing of a radial fin design is shown in Figure 4.22. If the fins are close together, the air flow velocity is low giving a lower convective heat transfer coefficient. So even though there is a large surface area the motor will run hotter than with a larger fin spacing with less total surface area. If the fin spacing is too large, the surface area is lost even though there is a good

Figure 4.21   Housing types commonly used with TENV cooling

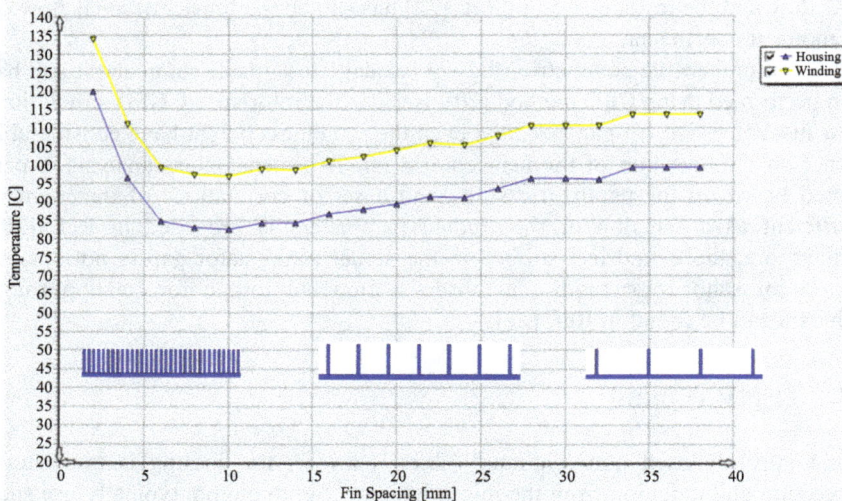

Figure 4.22   The variation of fin spacing with housing temperature and also the relationship between the housing temperature and winding temperature

amount of air flow and the temperature will again rise. In the design example, a spacing of around 10 mm between 2 mm fins is found to be optimal. The fin spacing optimization is calculated using the convection correlation built into Motor-CAD software in this case. CFD could also have been used to calculate the optimum.

Natural convection from a motor housing in air is typically between 5 and 10 $W/m^2/K$. If the housing has an emissivity of radiation above 0.9, which is typical for a painted surface, the radiation heat transfer coefficient will be of a similar magnitude or often larger than the convection heat transfer coefficient. Thus radiation will form an important component of motor cooling for TENV machines. The addition of fins will, however, increase the convection from the motor but will have a much smaller influence of the radiation from the housing, i.e. the effective radiation area for the housing being roughly equal to the motor outer envelope area.

The motor mounting arrangement can also have a significant impact on the thermal behavior of the motor. Often servo motors are rated when flange mounted to a large vertical metallic plate. Typically, for small motors, 50% or more of the total loss is dissipated through the flange. Such high figures are not uncommon as man-ufacturers tend to use larger cooling plates than the standard plate sizes recom-mended by NEMA [25] to create rating data published in catalogues. A motor must be de-rated if it is unable to dissipate such powers into the device it is mounted to.

A fin efficiency quantity is used to account for the reduction in fin temperature from fin base to tip due to conduction along the fin. The fin efficiency is the ratio of the actual heat transfer rate of the fin to the heat transfer of the fin if the entire fin temperature is at the fin base temperature. When radial fins are used to enhance the convective cooling, the fin efficiency will typically be close to 1. This is because in most cases, the fins are quite thick for manufacturability and handling safety, are not very long and are made from a high thermal conductivity material (often made from aluminum). In some special applications, thinner fins may be used but require protection from damage. Some motor users do not like to use fins to enhance cooling due to the added complexity of keeping them clean.

Servo motors are commonly used for motion control applications where they have an intermittent load. In such cases, the motors thermal performance is influ-enced by thermal capacitance of the different components as well the cooling of the housing. Figure 4.23 shows the predicted thermal transient for the housing and winding for two load situations. The first is for a continuous load. The second is a repeated intermittent load with a short period of three times the torque of the first case followed by a longer off period. The rms torque in both cases is the same. It is seen that the winding heats up very rapidly when overloaded due to its relatively small thermal capacitance compared to the bulk of the motor.

4.3 TEFC cooling

The majority of industrial induction motors designed for mains supply use the total enclosed fan-cooled (TEFC) cooling method where a fan is attached onto the shaft at the non-drive end of the motor. A fan cover or cowling is used to direct the fan air

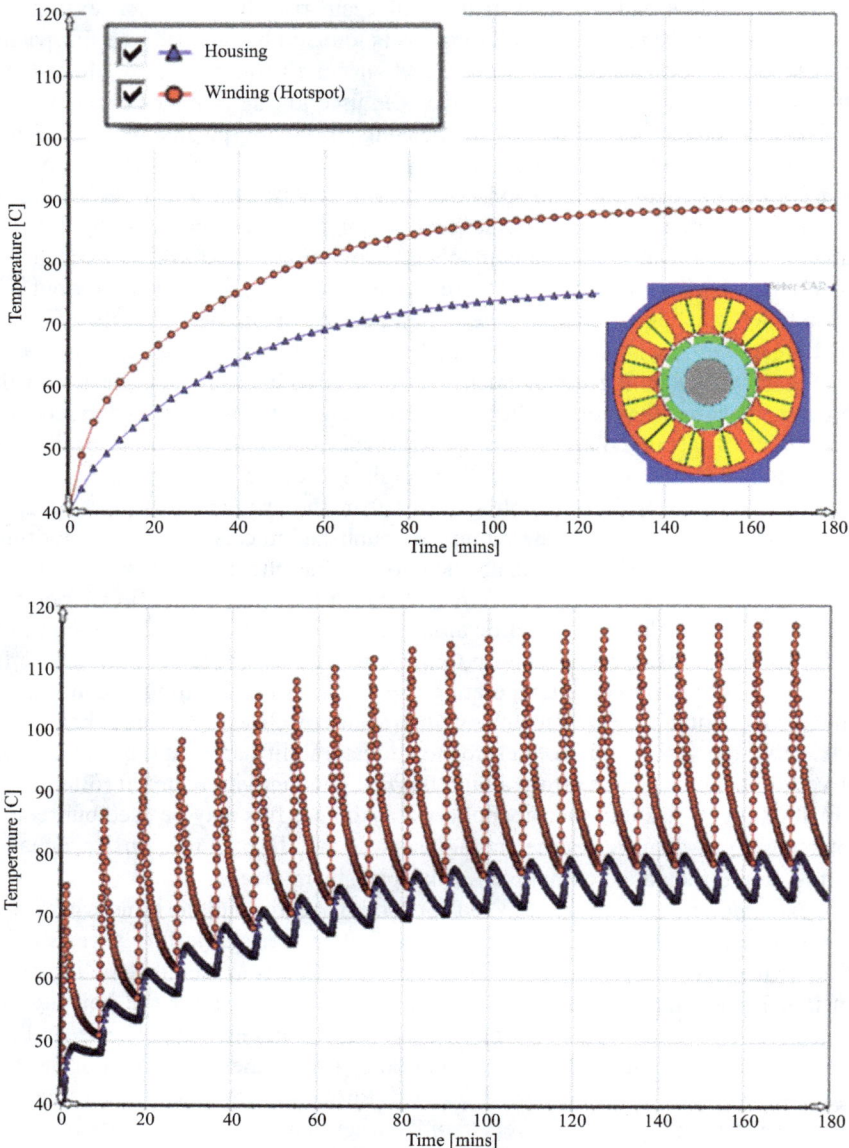

Figure 4.23   Thermal transient of a servo motor operating continuously and with a repetitive duty cycle load

flow to pass down axial fins channels on the outside of the housing. A typical arrangement is shown in Figure 4.24. Figure 4.25 shows some typical axial fin arrangements used for TEFC housings. The convective cooling is calculated using a correlation developed by Heiles [26] by performing many tests on open fin channels.

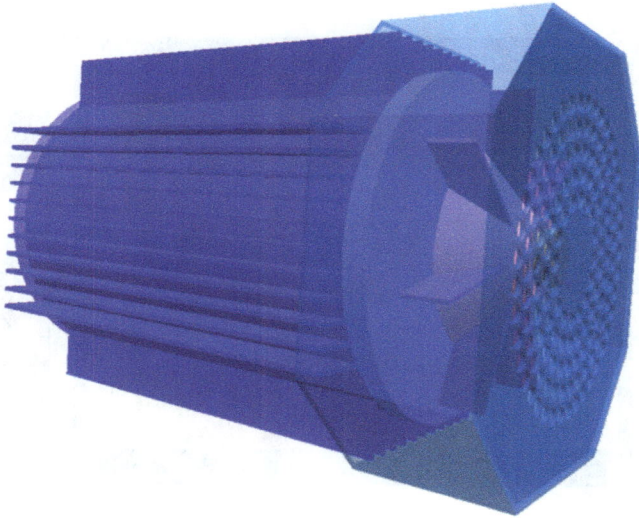

Figure 4.24   Typical TEFC housing, shaft mounted fan and cowling

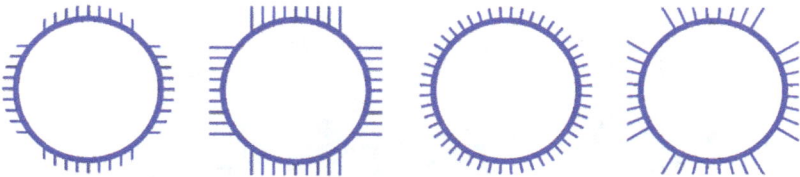

Figure 4.25   Housing types commonly used with TEFC cooling

Typically the fan cover or cowling does not extend over the full axial length of the motor housing. This will give lower pressure drop so providing a larger flow rate. It does however lead to a certain amount of leakage of air from the open fin channels. The amount of leakage increases with axial distance from the fan cover. An increase in air leakage means that the housing fin channel air velocity is lower at the motor drive end than at the non-drive end. The reduction in air velocity is indicated by the smaller flow arrows at the motor drive end in Figure 4.26. The typical form of the reduction in local fin channel air velocity is shown in Figure 4.27 [10]. The prediction of the actual reduction in velocity is a complex function of many factors including The fan design, fin and cowling design, and rotational speed. In the Motor-CAD software, the default amount of air leakage is set to be an average of the empirical characteristics shown in Figure 4.27. The designer can also easily set more or less leakage and so gain an understanding of the influence of the air leakage on the motor cooling. Once a promising housing design has been identified, a more accurate model can be developed if calibration is performed using testing and/or CFD.

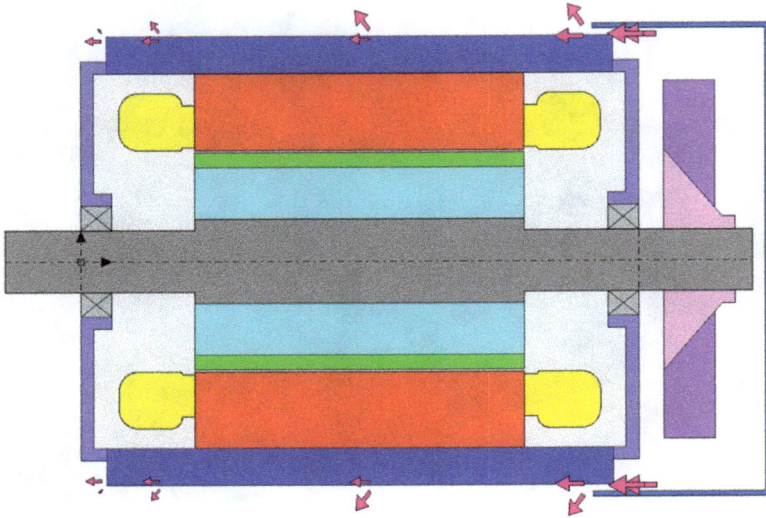

Figure 4.26   TEFC motor with flow arrows showing leakage of air from the open fin channels

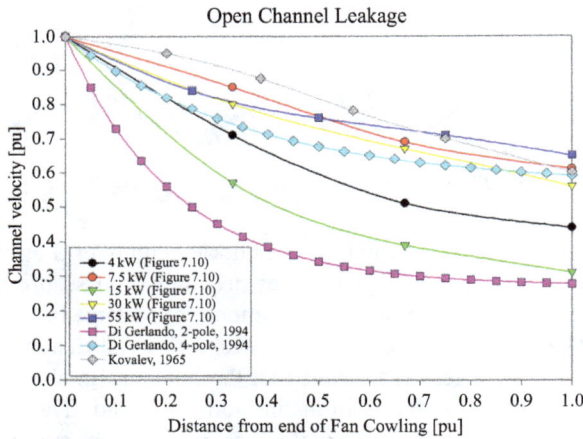

Figure 4.27   Typical form reduction in local fin channel air velocity with distance from the fan [10]

Not only the air velocity varies along the axial length of the open fin channels, but it often varies from one channel to another around the periphery of the housing as discussed more in Section 7.3. The variation is due rotational air flow effects and local blockage of the air flow by terminal boxes, lifting lugs, fan cover to housing connections, etc.

Housings with axial fins are usually cooled with a forced axial airflow. However, in some cases, the predominant cooling could be by natural convection and radiation. For instance, for a motor with a variable speed drive and a shaft mounted fan at or near to zero speed, the air flow from the fan will be non-existent or small. For a simple smooth cylindrical and square housings with no axial fins, the calculation of the natural convection and radiation heat transfer coefficients is quite simple. In such cases, it can be calculated using standard well-proven correlations for cylinders and flat plates with a radiation view factor of 1. The calculation of natural convection heat transfer coefficient and radiation view factor from an axially finned housing is more complex. In such cases, composite correlations based on all the geometric features of the housing are required for an accurate computation of the natural convection. When correctly used, this approach allows a good match between measured and predicted R_0 values for quite complex housing types [27]. These composite correlations are typically "area based" averaging existing correlations of the different parts of the housing (i.e. cylinder correlation for the fin channel base, horizontal flat plate for fins on the side of an axial finned housing, vertical U-shaped channel correlation for the fin on the other side of a radial finned housing, horizontal U-shaped channel for the fins on the top of an axial or radial finned housing, etc.). Special terms are required in the correlation to account for stagnant airflow in deep channels that have an orientation not optimal for natural convection. A good thing about using "area based" composite correlations are that if one of the geometric features dominates the structure, then the results will be very similar to those calculated using a single correlation for the dominant shape alone (i.e. for a horizontal finned hosing with very small fins the correlation will be mainly based on that of the horizontal cylinder). Radiation thermal exchange from finned housings is more complex than from smooth housings because of the more complex calculation of the view factor. As validated by several experimental tests, a correct use of complex composite convection correlations and view factor predictions allows excellent prediction of the convection thermal resistance between motor frame and ambient as shown in Figure 4.28. In this case, R_0 is the

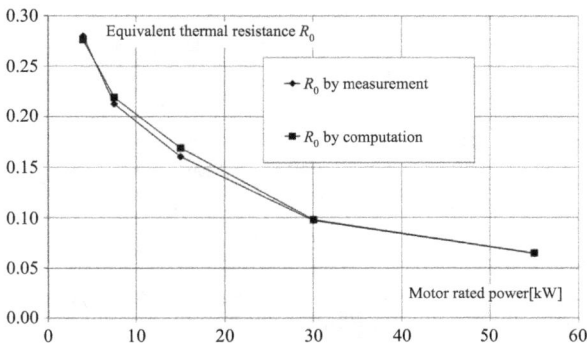

Figure 4.28   Comparison between the measured and the computed housing to ambient thermal resistance (R_0) from a DC calibration test with full load current for the five motors shown in Section 7.3

thermal resistance from the housing to ambient with a comparison of a composite correlation calculation with that from a DC calibration test performed at full load current for the five motors shown in Section 7.3.

The mixed heat transfer due to the combination of natural and forced convection can be estimated using the formula:

$$h^3_{mixed} = h^3_{forced} \pm h^3_{natural} \tag{4.24}$$

where the motor orientation determines the $\pm$ sign used, a $+$ for assisting and transverse flow and a $-$ sign for opposing flows.

In some applications, audible noise due to the ventilation air flow must be limited. In such cases, typically the air velocity should not exceed 20 m/s.

When axial fins are used to enhance convective cooling, the fin efficiency will usually be greater than 0.9. As with the radial fins used for TENV cooling, it is not practical to make the fins very thin and long. They are also most often made from a metallic material with a relatively high thermal conductivity (often made of cast iron or aluminum). The fin surface heat transfer coefficient will be significantly larger than that for TENV housing.

4.4   Open ventilated cooling

For totally-enclosed machines, the air temperature inside is typically high as it isolated from ambient, so it is not favorable for machine cooling. Compared to TENV and TEFC, open ventilated cooling is much better because ambient air is drawn into the electrical machine through filters and it has direct contact with the heat generating components such as the stator and rotor windings, magnets, rotor cage, etc. Due to the cooling effectiveness, open ventilated cooling is used in a range of applications including transportation systems (aerospace, railway, and marine propulsion), renewable energy (e.g. wind power generators), and industry (e.g. pumps and fans, manufacturing machines, and power generation).

Commonly, an open ventilated machine has a fan to force ambient air to pass through the air passages (cooling ducts). The fan can be shaft-mounted (open self-ventilated) or driven externally (open forced ventilated). For a shaft-mounted fan, the air flow rate is restricted by the machine speed and therefore it is not suitable for machines with variable speed operation, e.g. machines with high losses at low speed operation. Also, a shaft-mounted fan leads to additional fan windage loss which may unnecessarily reduce efficiency if the machine is not highly loaded. As an alternative solution, an externally powered fan can be used to generate the required cooling irrespectively of the machine rotational speed, as the fan speed can be controlled separately. The fan laws in Section 3.3.2 of Chapter 3 can be employed here to analyze the mechanical power required by either a shaft-mounted fan or an externally-powered fan.

To increase the benefit from direct contact between the cooling air and heat sources, cooling ducts are typically placed in the stator laminations, rotor laminations, and slots. For electrical machines with tooth wound coils (concentrated winding), an air void is formed between the coils as shown in Figure 4.29. This void is available as an air passage. As the cooling air is next to the coils, this can be effective for removing the heat due to resistive heating. An example of cooling ducts between tooth wound coils is given in Ref. [28] in which a liquid-flooded machine is described. For electrical machines with distributed multi-stranded winding, there is normally no extra space available for a cooling duct, or it will reduce the copper slot fill. Cooling ducts can be introduced in stator and rotor laminations as shown in Figure 4.30(a), but they must be accounted for in the

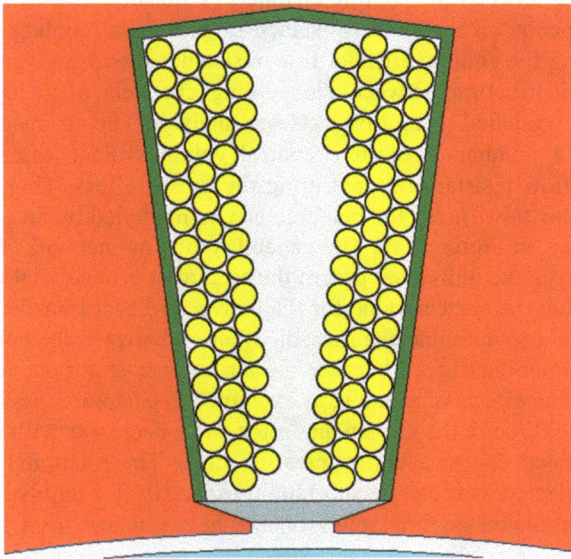

Figure 4.29   The air void formed between tooth wound coils

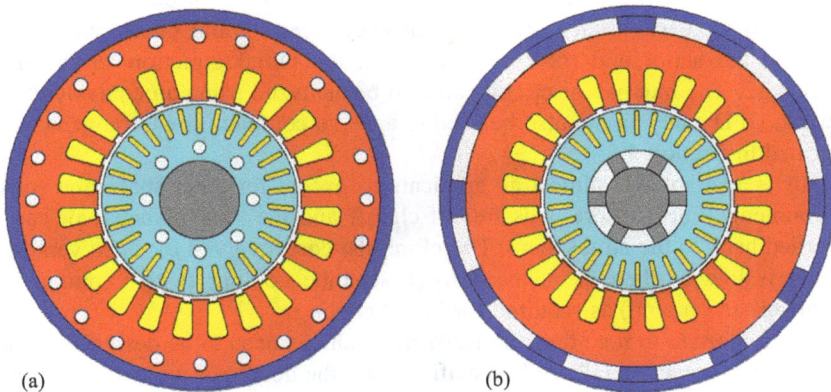

(a)   (b)

Figure 4.30   (a) An open ventilated machine with cooling ducts in stator and rotor laminations and (b) an open ventilated machine with cooling ducts in the housing and shaft

electromagnetic design of the back iron. Alternatively, the cooling ducts can be placed in the housing and shaft as shown in Figure 4.30(b).

For open ventilated cooling, the air flow rate is one of the major factors that affects cooling performance – the convective heat transfer is as a function of the fluid flow. However, as explained in Chapter 3, the air flow rate passing through an open ventilated machine is not only affected by the type of fan used but also the number of cooling ducts, duct geometry, their size, and location. Furthermore, the flow distribution between the cooling channels (stator ducts, rotor–stator gap, and rotor ducts) depends on the flow resistance of individual cooling channels. The factors that affect the cooling channel flow resistance are also given in Chapter 3.

The flow distribution between the cooling channels of an open ventilated machine can be modelled using flow network analysis. The air inside the machine is divided into a number of discrete control volumes. Each discrete volume is modelled by a flow resistance representing the pressure loss. Thus, the complete behavior of the air flow inside the machine can be modelled by an equivalent flow network as shown in Figure 4.31. The capability of flow networks for calculating the local velocity in the individual part of the air circuit is beneficial for the thermal model to calculate the correct value for the convective heat transfer coefficient.

It is important to note that the flow distribution between the cooling channels varies with the rotational speed because the air in rotor–stator gap and rotor ducts are subjected to the effects of rotation, as explained in Chapter 3. As a result, the air flow rate passing through the air gap and rotor ducts decreases with the increase in the rotational speed due to a higher pressure drop. The rotational pressure drop must not be neglected as it can be substantial especially for high-speed operation.

As open ventilated cooling relies on the heat transfer from the convective surface area and the cooling air, therefore it can be improved by increasing the convective surface area. This is achievable by having more cooling ducts and larger cooling ducts. However, the optimization of cooling channels has to be done by taking the machine's structural integrity, noise, and vibration aspects into account. In order to increase surface area for heat exchange, cooling ducts can even be placed in the stator and rotor core packs in the radial direction as shown in Figure 4.32. The internal air circuit can also be analyzed using an equivalent flow network. CFD analysis can also be used to analyze complex flow arrangements as described in Section 5.2.

Air is available for almost all applications except for outer space and subsea environments. Furthermore, it is free of charge and the "hot" exhaust air can be discarded back to the atmosphere. Therefore, the cost involved for open ventilated cooling is low as there is no need for a separate cooling system (pump, heat exchanger, and piping) as required for liquid cooling.

It is important to note that all electrical machines need to be designed to meet their Ingress Protection (IP) code specification – the degree of protection provided by electrical equipment enclosures against accidental direct contact with live parts and against the ingress of solid foreign objects or water. For example, IP22 machine is protected against objects greater than 12.5 mm in diameter such as finger contact and it is protected against vertical dripping water when the enclosure

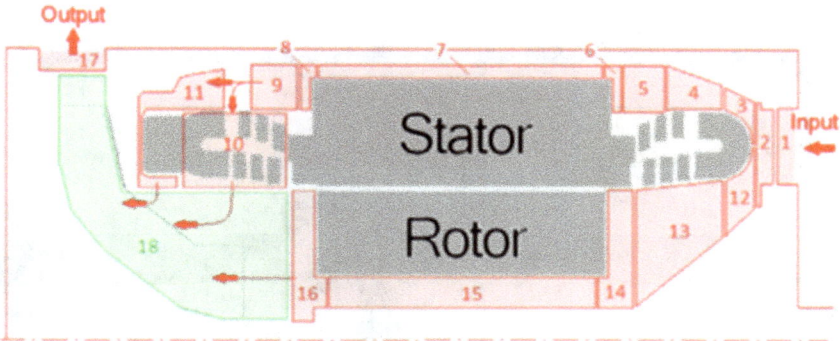

(a)

Number	Control volume	Type of resistance
1	Air input 1	Sudden contraction
2	Air input 2	Sudden expansion
3	End winding-frame 1	Smooth contraction
4	End winding-frame 2	Smooth expansion
5	End winding-frame 3	Annular straight length
6	Stator ducts entrance	Sudden contraction
7	Stator ducts	Rectangular straight length
8	Stator ducts exit	Sudden expansion
9	End winding-frame 4	Rectangular straight length
10	End winding	Series sum of 3 resistances[a]
11	End winding-frame 5	Series sum of 7 resistances[b]
12	End winding-end cap 1	Smooth contraction
13	End winding-end cap 2	Smooth expansion
14	Rotor ducts entrance	Sudden contraction
15	Rotor ducts	Circular straight length
16	Rotor ducts exit	Sudden expansion
17	Air output	Sudden expansion
18	Fan	—

[a] Two 90° bends and a straight length.
[b] Two contractions, two bends, two straight lengths and one expansion.

(b)

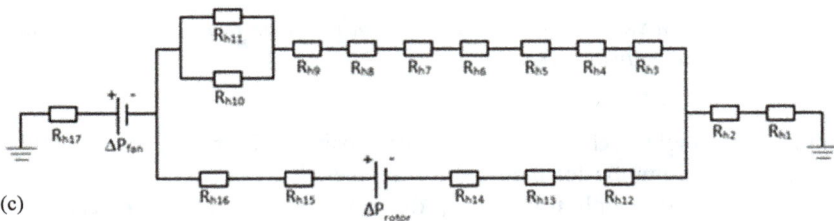

(c)

Figure 4.31 (a) Discretization of air volume inside an open ventilated machine, (b) control volumes identified in (a), and (c) its equivalent flow circuit [29]

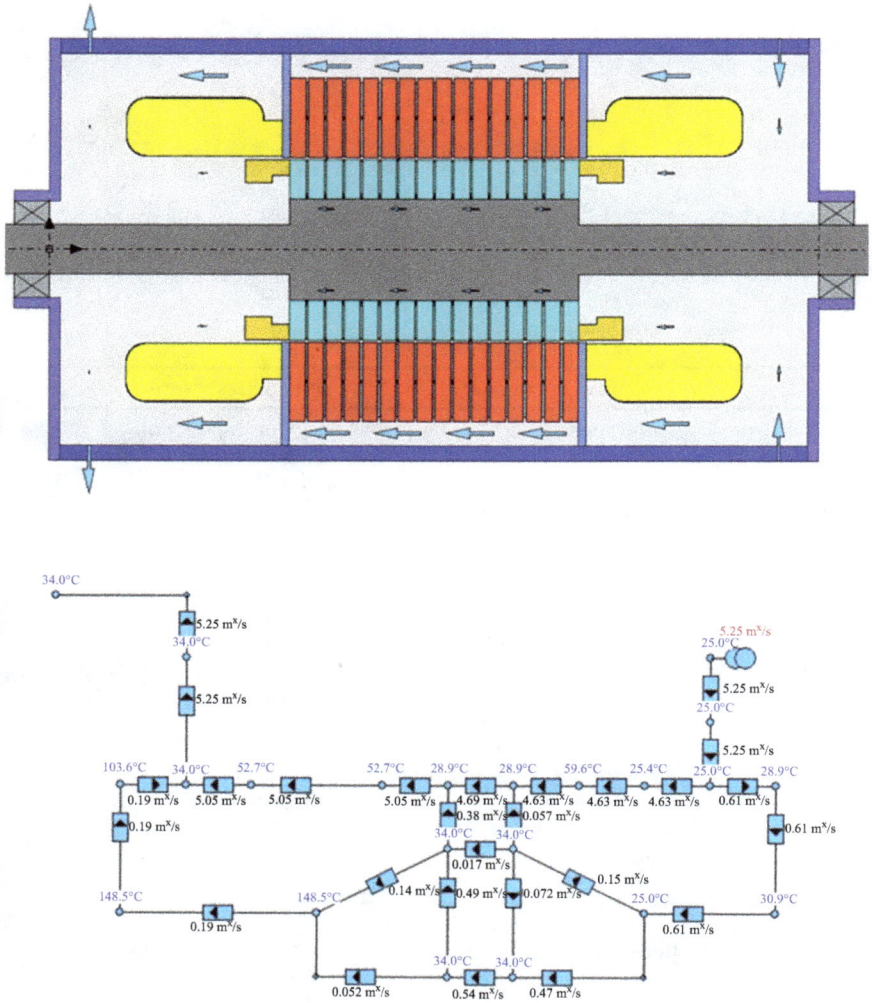

Figure 4.32   Open ventilated machine with radial ducts (top) and its equivalent flow network (bottom)

is tilted at an angle of 15° from its normal position. Therefore, open ventilated cooling is quite common for indoor use machines. For operation in dirty environment, filters are commonly used to stop dirt from entering into the machine.

For outdoor use, machines are normally protected by angled louvres and filters, which are suitable for most environments. In areas subjected to heavy rainfall, hoods are fitted to protect the air inlets and exhausts. Inertial separators can be fitted at the air inlets to remove excessive sand or large dust particles. In marine environments, coalescing filters can be used to remove salt laden moisture droplets.

However, filters need regular maintenance and a blocked filter could cause an additional pressure drop leading to a reduction in the air flow rate.

The cooling system design for through ventilated electrical machines can be undertaken using computational fluid dynamics in which the fan and cooling system within the machine can be modelled and optimized. This is described more fully in Section 5.2.

4.5 Close circuit cooling with heat exchanger

Closed circuit cooling has been widely used for different types of turbo generator in the power industry, such as wind, hydro, gas, and steam turbines, that can produce power of multi-megawatts. The stator and rotor cores of turbo generators normally have radial ventilating ducts at intervals along the cores as illustrated in Figure 4.33. The radial ducts allow more uniform cooling of the winding as the windings are closer to the cooling air and this reduces the conduction thermal resistance. The radial duct configuration significantly increases the surface area for convective heat transfer especially for large turbo generators. As turbo generators provide power in sectors with the technically demanding and hazardous operating environments such as on offshore platforms and petrochemical plants, they normally have to be designed for IP of IP55 or IP56. As a result, air is forced around the generator by means of axial fan mounted on the rotor shaft and the hot exhaust air is cooled by heat exchanger located on the top of the generator before being recirculated back to the machine inlet.

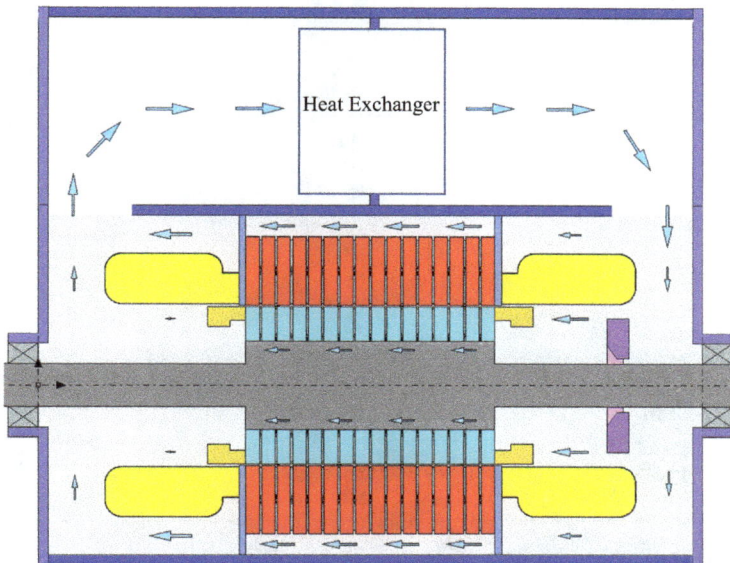

Figure 4.33   A typical closed circuit-cooled turbo generator with heat exchanger

Since the rotor–stator gap flow resistance is relatively high, the rotor cooling air also mainly supplies the stator cooling air. To avoid the rotor cooling air from bypassing the radial ducts, the other end of the rotor axial ducts is blocked. Sometimes the stator core is too long and it is challenging to supply sufficient air to the stator core through the rotor–stator gap and the rotor ducts. This is overcome by having inlets at both ends of the machine as shown in Figure 4.34, rather than the single inlet configuration as shown in Figure 4.33. The outlet is over the stator core. Cooling air is driven by two shaft-mounted fans at the drive and non-drive ends, respectively. This ensures sufficient air flow to the entire core length.

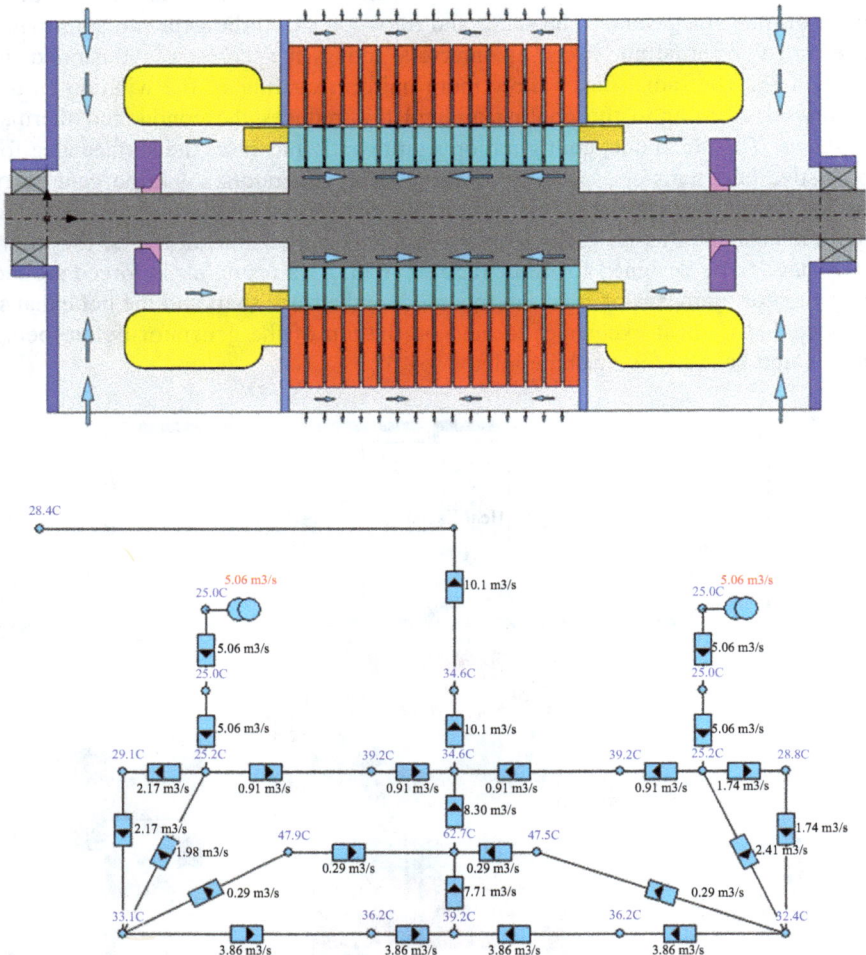

Figure 4.34   Close circuit-cooled machine with inlets at both ends (top) and its equivalent flow network (bottom)

Table 4.7   Comparison of the thermal properties of air and hydrogen

Thermal properties at 25 °C		
Thermal properties	**Air**	**Hydrogen**
Density, ρ (kg/m^3)	1.18	0.083
Thermal conductivity, k [W/(m.K)]	0.026	0.181
Specific heat capacity, c [kJ/(kg K)]	1.01	14.4
Dynamic viscosity, μ [kg/(m s)]	1.83×10^{-5}	8.9×10^{-6}
Thermal properties at 100 °C		
Thermal properties	**Air**	**Hydrogen**
Density, ρ [kg/m^3]	0.946	0.066
Thermal conductivity, k [W/(m K)]	0.032	0.215
Specific heat capacity, c [kJ/(kg K)]	1.01	14.5
Dynamic viscosity, μ [kg/(m s)]	2.18×10^{-5}	1.04×10^{-5}

The air temperature inside a closed circuit machine is an important variable that affects the overall cooling performance of the machine. Unlike open ventilated machine, the cooling air temperature of a close circuit-cooled machine is entirely determined by the heat exchanger. Normally the heat exchanger is either air or water cooled and the choice is dependent on which is more suitable for the machine application. For example, due to lack of water in Middle Eastern and African countries, air-cooled heat exchanger is more likely. The selected heat exchanger must be able to dissipate the losses generated in the machine.

An alternative coolant to air, hydrogen gas, is commonly used for turbo generators due to its favorable thermal properties. As shown in Table 4.7, hydrogen is about seven times more conductive than air and has higher specific heat capacity. Moreover, hydrogen is less viscous and has much lower density. For turbo generators, due to the radial ducts and impeller in the rotor, the windage loss is usually substantial, so a higher fluid flow rate can be induced by the impeller. Therefore, by using hydrogen as the coolant, the frictional drag of the rotor can be reduced considerably. Consequently, hydrogen-cooled turbo generators give higher output power and efficiency compared to air-cooled machines. However, there is a safety risk because hydrogen is flammable when mixed with air. To avoid that, the purity of the hydrogen coolant is crucial which is kept above 90% [30,31].

4.6   Housing water jacket cooling

A housing water jacket is a favorable cooling method widely used for high-performance electrical machines, where the size and the weight are critical. This cooling method is also widely used with other liquid coolant types in spite of it being referred to as a water jacket. For instances, ethylene glycol-based water solutions are commonly used as the coolant in automotive traction motors. Aviation turbine fuel is commonly used as the coolant of aircraft generators.

Compared to air, the thermal properties of liquids result in a much higher convective heat transfer coefficient. Table 4.8 shows a comparison between water and air based on the same Reynolds number. With a Reynolds number of 8×10^3 and a duct diameter of 10 mm, the heat transfer coefficient for water is about 58 times that of air. However, pure water cannot be used as a coolant in some applications as it freezes in cold weather and normally it is mixed with antifreeze (e.g. ethylene glycol) to keep it from freezing. Ethylene glycol 50% is a common automotive coolant used in housing water jackets. Table 4.9 shows the comparison between ethylene glycol 50% and water. As ethylene glycol 50% is more viscous than water, based on the same fluid velocity the resulting Reynolds number of ethylene glycol 50% is about a quarter of Reynolds number of water. As a result, the convective heat transfer coefficient of ethylene glycol 50% is only 0.43 of heat transfer coefficient of water. On the other hand, the pressure requirement for both ethylene glycol 50% and water are very similar.

By using housing water jacket cooling, an electrical machine can be fully sealed so a high level of safety can be ensured (i.e. IP68), protected from dust and protected against immersion in water. Commonly, the housing water jacket is formed between two pieces of metal, an inner housing and an outer housing. As illustrated in Figure 4.35, the cooling channel grooves are usually created in one of the housings.

Table 4.8 *Comparison between the cooling properties of water and air for* $Re = 8 \times 10^3$ *and a duct diameter of 10 mm (the calculation is based on fluid properties at 20 °C)*

Cooling	Water	Air	Water/air ratio
Nusselt number, Nu	66.6	26.6	2.51
Heat transfer coefficient, h [W/(m^2 K)]	3954	68.2	58
Mean velocity, V [m/s]	0.80	12.0	0.067
Pressure drop, Δp [Pa]	$K\rho_w V_w^2/2$	$K\rho_{air} V_{air}^2/2$	$\frac{\rho_w V_w^2}{\rho_{air} V_{air}^2} = 3.7$

Table 4.9 *Comparison between the cooling properties of ethylene glycol 50% and water for duct fluid velocity of 2 m/s and duct diameter of 10 mm (the calculation is based on fluid properties at 20 °C)*

Cooling	Ethylene glycol 50%	Water	Ethylene glycol 50%/water ratio
Reynolds number, Re	5,357	19,924	0.27
Nusselt number, Nu	90.6	138.3	0.66
Heat transfer coefficient, h [W/(m^2 K)]	3,498	8,205	0.43
Pressure drop, Δp (Pa)	$K\rho_w V_w^2/2$	$K\rho_{air} V_{air}^2/2$	$\frac{\rho_w V_w^2}{\rho_{air} V_{air}^2} = 1.08$

(a)   cooling channel grooves in inner housing

(b)   cooling channel grooves in outer housing

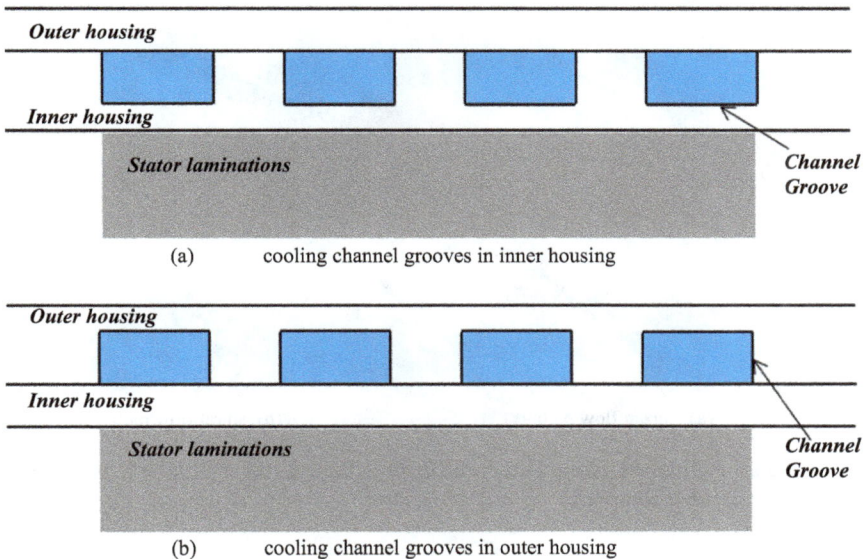

Figure 4.35   Housing water jacket with cooling channel grooves in (a) inner housing and (b) outer housing, respectively

As the effectiveness of housing water jacket cooling is dependent on the temperature difference between the housing and the liquid coolant, the housing is usually made of thermally conductive metals, i.e. aluminum or non-magnetic steels. Due to the interface gap between inner and outer housing, the conduction heat transfer is reduced, so, the channel walls of inner housing are usually considered to be the primary convective surface area. In addition, the channel wall thickness is an important variable that affects the temperature distribution in the radial direction. A small wall thickness leads to high-temperature gradient and hence the effective primary convective surface area reduces. On the other hand, large wall thickness gives a lower temperature gradient but the number of cooling channels, and hence surface area for convection, that can provided within the limited space reduces. Consequently, channel wall thickness needs to be optimized to achieve the best cooling performance.

Besides the convective surface area and the temperature difference between the channel wall and liquid coolant, another parameter that influences housing water jacket cooling is the convective heat transfer coefficient. Based on the lumped parameter thermal network method, the heat transfer coefficient can be determined from the internal flow heat transfer correlations presented in Chapter 2. Since the shape of cooling channel cross-section is normally rectangular, hydraulic diameter can be used in the correlations.

In housing water jacket designs, there are two typical flow paths – zigzag flow and spiral flow as shown in Figure 4.36. For the zigzag design, the flow moves

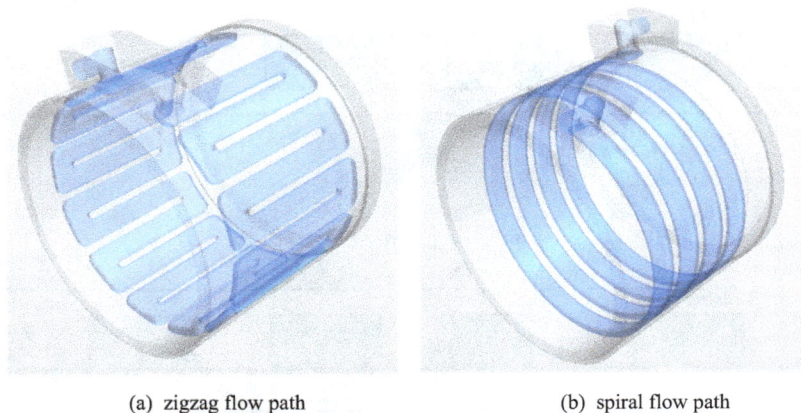

(a) zigzag flow path   (b) spiral flow path

Figure 4.36   Typical housing water jacket flow paths: (a) zigzag flow path (b) spiral flow path

forwards and backwards axially through bends around the housing. The changes in flow direction can lead to higher convective cooling compared to spiral flow. In the spiral flow design, the flow moves around the housing through a spiral channel. The pressure drop of spiral flow is mainly due to duct wall friction, however, for zigzag flow, the pressure drop is higher for a given flow rate due to the changes in flow direction at each bend. If the flow rate is not sufficiently high, the fluid temperature rise can be significant and this can lead to an asymmetric cooling effect depending on the flow path pattern. Zigzag flow can result in uneven cooling in the circumferential direction and therefore may not suitable for machines of large diameter. Spiral flow results in uneven cooling in the axial direction and hence may not be suitable for machines of long stack length. The significance of asymmetric cooling effect can be analyzed from the fluid temperature rise which can be predicted based on the amount of heat to be dissipated and fluid thermal properties.

To address the pressure drop issue, it is common for the cooling channels of housing water jackets to be connected in parallel. For instance, the zigzag flow design in Figure 4.36(a) can be replaced by two inlets and two outlets. Attention must also be paid to the design of the inlet and outlet manifolds as inappropriate design may result in a high pressure drop and a higher flow resistance will reduce the flow rate from the pump. With a parallel connection, for a given total flow rate, the flow velocity in each flow path is lower. Further analysis is then required to investigate the impact of change of flow velocity on convective heat transfer coefficient.

For an electrical machine that is mainly cooled by a housing water jacket, the heat generated in the windings and stator iron has to transfer through the interface gap between stator and housing as shown in Figure 4.37(a). Therefore, the contact resistance between the stator and the housing is very critical and it affects overall

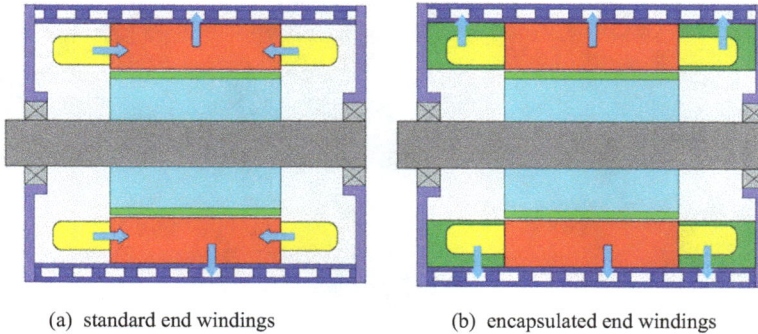

(a) standard end windings   (b) encapsulated end windings

Figure 4.37   Housing water jacket-cooled machine with end windings
encapsulated using potting material: (a) standard end windings and
(b) encapsulated end windings

machine cooling. Several empirical studies have been done in Refs [16–18], to provide information about the possible range of thermal contact resistance and the main factors that affect it.

For certain machines, the copper loss in the end windings due to joule effect can be substantial. The heat flow path for this end winding loss is relatively long compared to the active part of the winding as shown in Figure 4.37(a). Furthermore, the end windings are surrounded by the internal air which may be warm. This may not provide effective convective cooling to the end windings and consequently, the machine hot spot may be located in the end windings. In order to shorten this heat flow path and reduce the thermal resistance, several studies [32,33] have proposed to fill the space between the housing and end windings with high thermal conductivity encapsulation material as shown in Figure 4.37(b). The common encapsulation materials suppliers are LORD [34], Elantas, Huntsman, etc. Their thermal conductivities are typical in the range of 0.2–1.5 W/m/K and some encapsulation materials can have an even higher thermal conductivity, although it is worth noting that encapsulation materials with higher thermal conductivity are normally denser and more viscous. By encapsulating end windings, the experimental investigation presented in Ref. [32] has demonstrated that the machine hot spot temperature can be reduced considerably. Therefore, the winding current density can be increased whilst the machine hot spot temperature remains within the thermal limits.

An example of sensitivity analysis to the thermal conductivity (0.1–4 W/m/K) of the encapsulation material in an induction machine using Motor-CAD software is shown in Figure 4.38. With higher thermal conductivity material, the analysis shows that the temperature difference between the winding and the housing reduces enhancing the overall machine cooling performance.

Besides housing cooling jacket, direct cooling at the end-windings was proposed by inserting dedicated liquid cooling pipe in the end windings as shown in Figure 4.39. The cooling system has two hydraulic circuits: water jacket, cooling pipe which can be connected in parallel. The proposed cooling can significantly

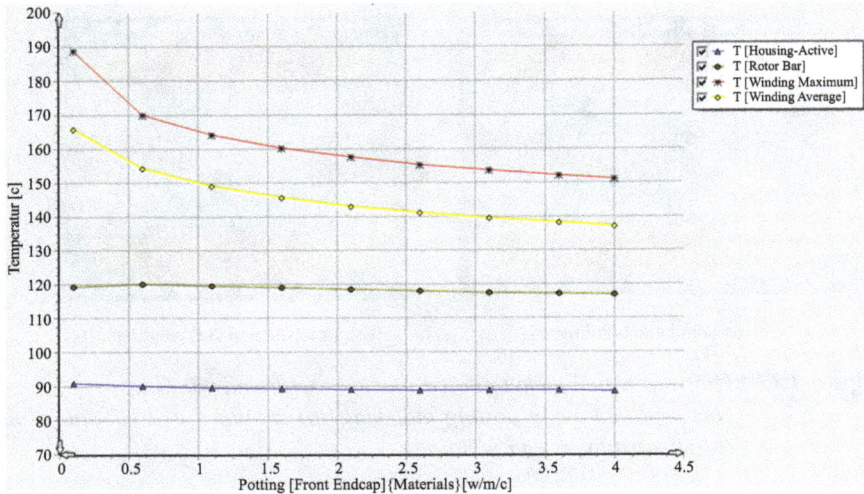

Figure 4.38   The variation of potting material thermal conductivity with machine temperatures

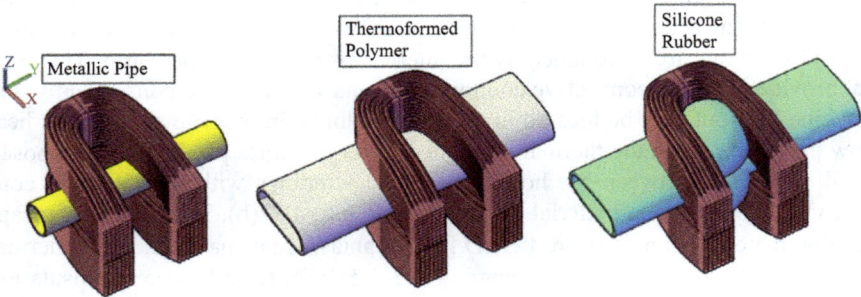

Figure 4.39   Direct end windings cooling for different materials [35]

reduce not only end-windings temperature but also winding temperature in the slots by taking advantage of excellent thermal conductivity of copper in the axial direction. Different pipe materials have been investigated as depicted in Figure 4.39, e.g. silicone rubber, thermoformed polymer (PTFE), metallic pipe. Their performances are compared in terms of cost, thickness, thermal conductivity, flexibility to ensure maximum contact area for heat exchange, non-magnetic material to avoid induced losses, high thermal limit, and chemical stability for desired machine lifetime. The comparison reveals that silicone rubber is most suitable material and overperform other materials in terms of cooling. As end-winding hot-spot and slot winding temperature are reduced, this leads to increment in machine current density.

A housing water jacket is a common cooling method for the stator, however, it may not provide sufficient cooling for the rotor. Therefore, it is only suitable for electrical machines with small losses in the rotor such as permanent magnet machines. For induction machines where there are high losses in the rotor due to the rotor cage, additional active cooling might be required to keep the rotor temperature below its thermal limits.

4.7   Sleeve with flooded stator cooling

In electrical machines where the heat is dissipated primarily through the frame, either with air cooling, as described above in Sections 4.2 and 4.3, or liquid jacket cooling, described above in Section 4.6, the main heat flow path for the stator copper and iron losses is conduction radially through the stator laminations. In high-power machines, the thermal resistance imposed by conduction through the stator may result in high winding temperatures limiting the machine power density. This is particularly true for larger machines, where the conduction path in the laminations is longer and hence thermal resistance higher. Introducing coolant closer to the sources of loss would reduce this thermal resistance. In low-power density machines, this can be achieved by using open air ventilation arrangements as described in Sections 4.4 and 4.5 above. However, the heat transfer coefficients achievable with air are typically one or two orders of magnitude lower than those achievable with liquid cooling, as shown in Table 4.8. More intensive cooling of higher power density machines could therefore be achieved with liquid cooling in closer contact with the sources of heat in the windings and laminations. This can be achieved by passing liquid coolant down axial ducts within the stator of a machine or down ducts within the stator windings themselves. Similarly, more intensive cooling of the end windings can be achieved by flooding the end region with a liquid coolant.

More intensive cooling of a machine could be achieved therefore if a liquid coolant were used to flood the machine. However, this would have the effect of increasing the windage drag on the rotor. In Section 4.1.7.7, Eq. (4.23) shows that windage loss in the rotor–stator annular gap varies linearly as density and (rotational speed)[3]. As liquids have densities typically two orders of magnitude greater than air and, in high-speed machines, a fully liquid-flooded machine is likely to have unacceptably high windage losses. For lower speed machines, the marginal increases in windage loss may be acceptable in liquid-flooded machines given the benefits in enhanced cooling and an oil spray cooling arrangement is described in Section 4.8 below. Ref. [28] also describes an oil-flooded machine in which the oil coolant is used to provide cooling in ducts in between the stator windings.

However, in high-speed permanent magnet machines, most of the loss in the machine (typically 80% or more) is generated in the stator and the loss in the rotor is small in comparison (typically less than 10%). In these machines, the main cooling requirement is for the stator with the rotor requiring much less cooling.

A means of achieving this is to use a sleeve in the rotor–stator gap to physically separate the stator from the rotor as shown in Figure 4.40. Liquid coolant can then be used to achieve intensive cooling through ducts on the inside and the outside of the stator, whilst the rotor, which requires proportionally little cooling, rotates in air. The sleeve prevents oil from entering the rotor–stator gap, avoiding excessive windage loss.

A case study for a machine with a sleeve and a liquid flooded stator is described in Chapter 7, Section 7.7.

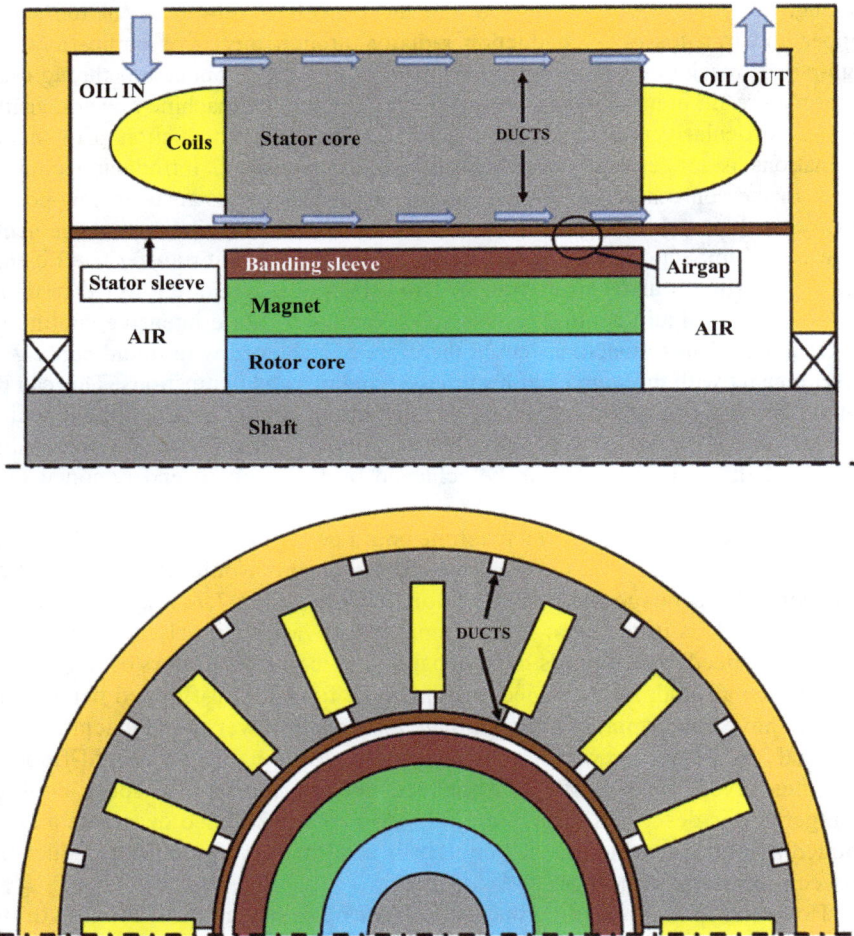

Figure 4.40   Cross-sectional views of the sleeved, flooded stator permanent magnet machine [36]

4.8 Oil spray cooling

In this section, oil spray cooling is not limited to oil spray only, it also includes oil jets and oil dripping cooling. By reviewing the cooling systems of modern and commercially available high-performance electrical machines for automotive applications [37,38,50], aerospace application [39], and the cooling of electronic chips [40]. Oil spray cooling is one of the main cooling systems to meet high-current density. Since the copper loss is the major loss component, end windings can be impinged directly by oil. This mitigates the thermal resistance between slot windings and housing for housing cooling jacket which is relatively high.

There are many spray nozzles available in the market that provide different spray patterns. Fundamentally, for oil spray cooling, oil is atomized through a spray nozzle and breaks into small droplets dispersing onto the cooled surface area. As spray nozzles convert pressure energy into kinetic energy, the pressure of the oil is an important factor to ensure good atomization. The shape of the spray is an indication of the degree of atomization. When compared to conventional housing cooling jacket, the oil atomization process involves higher pumping pressure. It is normally in the order of magnitude of a few bar. Based on Davin's investigation [41], two types of nozzles are used: a full cone nozzle (FCN) and a flat jet nozzle (FJN) each produces a different shape of spray. At lower flow rate, a continuous film of oil is formed (in either the hollow cone or flat jet) before it breaks up at the edges as shown in Figure 4.41. Beyond a certain flow rate, a finely atomized spray is formed as shown in Figure 4.41. The critical flow rate is found to be affected by oil temperature and is lower at higher oil temperature because the oil becomes less viscous.

Similar to oil spray, oil can be forced to pass through a set of orifices in an oil tube as shown in Ref. [41]. This creates oil jets or dripping oil targeting at the end windings depending on the pressurization level of the oil and gravitational effects. Due to gravity, spray nozzles/orifices tend to be located above the upper part of end windings. The lower part of end-winding is wetted and cooled by the falling oil film. Jet impingement cooling provides high heat transfer coefficient but they are not uniform over the cooled surface. As shown in Figure 4.42, the heat transfer coefficient is the highest at the stagnation zone (also known as impingement region) and decreases as the oil flows away from the stagnation zone. When jet impingement cooling is applied to electrical machine for end winding cooling, the non-uniform heat transfer coefficient causes asymmetric cooling effects, with the jet impinged surface areas subjected to higher heat transfer coefficients while the other end winding and wall surfaces are cooled by the falling liquid film of variable thickness. It is important to note that the heat transfer correlations [39,40,42] for jet impingement cooling are developed from ideal case with flat surfaces, whereas an actual end winding has a much more irregular surface. End windings with a high degree of openness, such as hairpin windings, allow the oil to penetrate into the end windings which increases the benefit of having oil spray cooling. However, there is a concern that high-speed impinging jets could damage the wire insulation systems and therefore care must be taken.

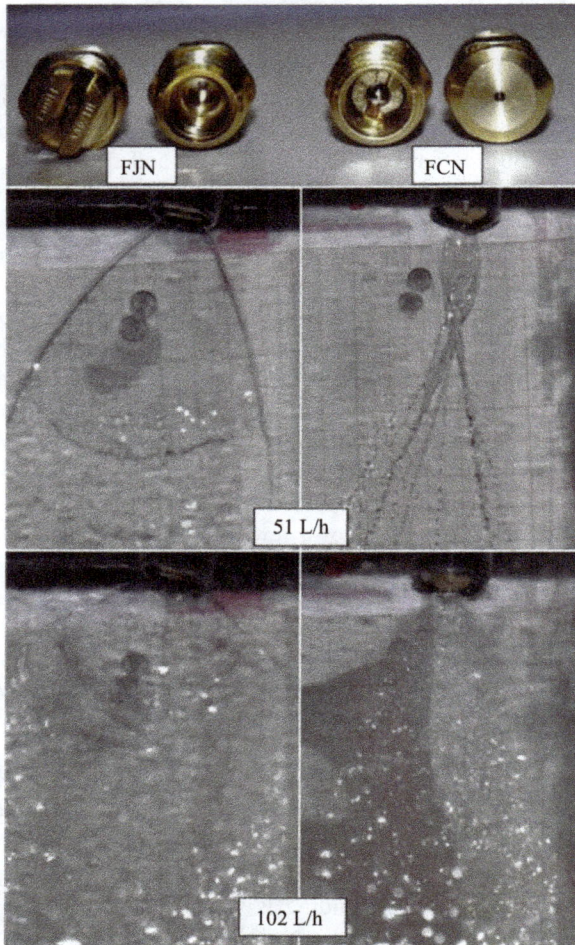

Figure 4.41   Oil flow pattern for the FJN and the FCN for low (top: 51 L/h) and high (bottom: 102 L/h) flow rates at 50 °C [41]

Despite the complexity of oil spray characteristics, the rotating rotor creates more turbulence and effectively mixes oil droplets with air in the end-space regions. This could further enhance the oil spray cooling and decrease the temperature variation of the entire end windings. Moreover, besides "static" oil spray described above, rotating oil spray was also used as proposed in Ref. [37], to fully exploit the oil spray cooling benefits. Oil is fed into the center of rotor core through hollow shaft. Then oil flows to rotor ends from center. This cooling structure not only cools the rotor core actively but also provides oil splashing cooling to the end winding inner surfaces.

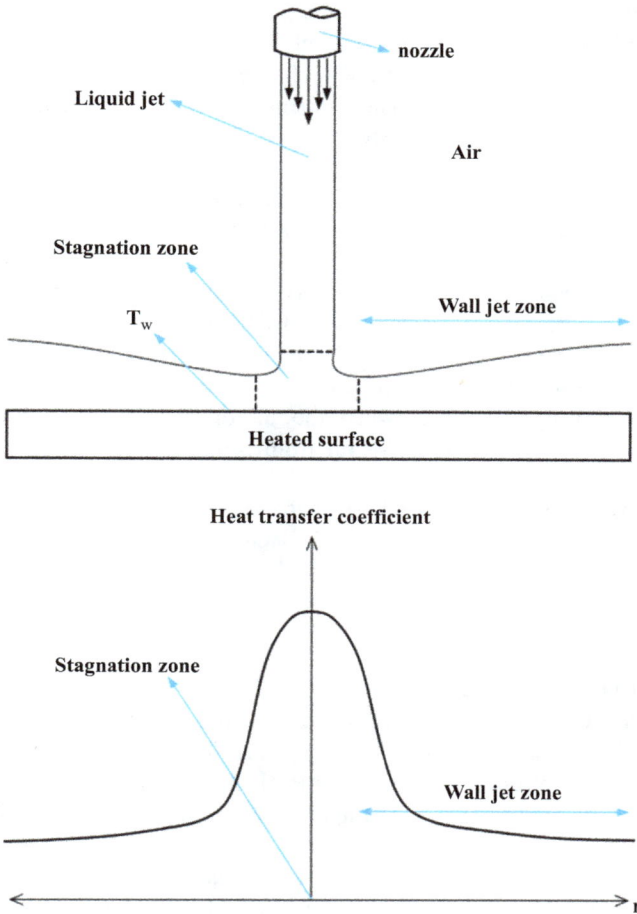

Figure 4.42   Schematic of jet impinging on a surface (top) and heat transfer coefficient variation along the cooled surface (bottom)

Due to the effectiveness of oil spray cooling, the HTC value of oil spray cooling is the main interest of thermal engineers. However, there are challenges in analyzing that using CFD method. A few direct experimental measurements have been performed. In Ref. [42], HTC of automotive transmission oil jet can reach up to about 10,000 W/m^2/K at jet velocity of 10 m/s on a fabricated test sample with target surface of 12.7 mm in diameter. Based on actual geometry of end windings, the averaged HTC values on end winding with the proposed setup in Ref. [43] ranges from 100 to 550 W/m^2/K. The main factors that affect oil spray cooling are oil velocity and rotor speed.

For oil spray cooling, the coolant used is normally already present in the system. For example, lubricating oil and jet fuel for aircraft electrical machines, automatic transmission fluid (ATF) and gearbox oil for automotive traction motors. However, the thermal properties of these oils are very sensitive to the temperature especially the viscosity. For Mercon LV ATF, the kinematic viscosity is 29.6 mm^2/s at 40 °C and 6 mm^2/s at 100 °C. The temperature-sensitive thermal properties not only affect the convective heat transfer rate but also the oil pressure required for a given flow rate and windage loss.

An additional drawback of oil spray cooling is the additional windage loss due to possibly intrusion of oil mist into the rotor–stator gap. By changing the cooling method of an induction machine from TEFC to oil spray cooling, the total mechanical loss can increase considerably, by about three to four times as demonstrated in Ref. [44]. In Ref. [38], "oil deflectors" have been demonstrated successfully to keep oil from getting into the rotor–stator gap. If the rotor is submerged under the oil, this will lead to even higher machine loss due to drag. Therefore, a scavenge pump between the oil drain hole and oil reservoir is recommended to withdraw excessive oil from filling the end space of oil spray-cooled machine.

An experimental test rig used to characterize different types of oil spray cooling performance is given in Chapter 7, Section 7.8.

References

[1]   Staton D. and Popescu M. 'Analytical thermal models for small induction motors'. *COMPEL – Int. J. Comput. Math. Electrical Electron. Eng.* 2010;29(5):1345–1360, https://doi.org/10.1108/03321641011061542

[2]   Simpson N., Wrobel R., and Mellor P. H. 'Estimation of equivalent thermal parameters of electrical windings'. *IEEE Trans. Ind. Appl.* 2013;49(6): 2505–2515.

[3]   Idoughi L., Mininger X., Bouillault F., Bernard L., and Hoang E. 'Thermal model with winding homogenization and FIT discretization for stator slot'. *IEEE Trans. Magn.* 2011;47(12):4822–4826.

[4]   Hashin Z. and Shtrikman S. 'A variational approach to the theory of the effective magnetic permeability of multiphase materials'. *J. Appl. Phys.* 1962;33(10):3125–3131.

[5]   Milton G. W. 'Bounds on the transport and optical properties of a two component composite material'. *J. Appl. Phys.* 1981;52:5294–5304.

[6]   Wrobel R. and Mellor P. H. 'A general cuboidal element for three-dimensional thermal modelling'. *IEEE Trans. Magn.* 2010;46(8):3197–3200.

[7]   Simpson N., Wrobel R., and Mellor P. H. 'A general arc-segment element for three-dimensional thermal modeling'. *IEEE Trans. Magn.* 2014;50(2): 265–268.

[8]   Schubert E. 'Heat transfer coefficients at end-winding and bearing covers of enclosed asynchronous machines'. *Elektrie* 1968;22:160–165.

[9]   Boglietti A. and Cavagnino A. 'Analysis of the endwinding cooling effects in TEFC induction motors'. *IEEE Trans. Ind. Appl.* 2007;43(5):1214–1222.

[10] Staton D., Boglietti A., and Cavagnino A. 'Solving the more difficult aspects of electric motor thermal analysis in small and medium size industrial induction motors'. *IEEE Trans. Energy Convers.* 2005;20(3):620–628.

[11] Basso G. L., Goss J., Chong Y. C., and Staton D. 'Improved thermal model for predicting end-windings heat transfer'. *IEEE Energy Conversion Congress and Exposition (ECCE)*, Cincinnati, *OH*, USA, October 2017, pp. 4650–4657.

[12] International Standards Office, ISO 15312: 2003. *(E) Rolling Bearings—Thermal Speed Rating—Calculation and Coefficients*, Geneva: ISO.

[13] German Standards, DIN 732:2010-05 *Rolling Bearings – Thermally Safe Operating Speed Calculation and Correction Values*, Berlin: DIN.

[14] SKF, Rolling Bearings, www.skf.com; 2020.

[15] Cousineau J. E., Bennion K., DeVoto D., and Narumanchi S. 'Experimental characterization and modeling of thermal resistance of electric machine lamination stacks'. *Int. J. Heat Mass Transfer* 2019;129:152–159.

[16] Kulkarni D. P., Rupertus G., and Chen E. 'Experimental investigation of contact resistance for water cooled jacket for electric motors and generators'. *IEEE Trans. Energy Convers.* 2012;27(1):204–210.

[17] Cousineau J. E., Bennion K., Chieduko V., and Gilbert A. 'Experimental characterization and modeling of thermal contact resistance of electric machine stator-to-cooling jacket interface under interference fit loading'. *J. Therm. Sci. Eng. Appl.* 2018;10:1–7.

[18] Simpson N., Wrobel R., Booker J. D., and Mellor P. H. 'Multi-physics experimental investigation into stator-housing contact interface'. *8th IET International Conference on Power Electronics, Machines and Drives*, Glasgow, UK; 2016, pp. 1–6.

[19] Boglietti A., Cavagnino A., and Staton D. 'Determination of critical parameters in electrical machine thermal models'. *IEEE Trans. Ind. Appl.* 2008;44(4):1150–1159.

[20] Nachouane A. B., Abdelli A., Friedrich G., and Vivier S., 'Estimation of windage losses inside very narrow air gaps of high speed electrical machines without an internal ventilation using CFD methods'. *International Conference on Electrical Machines (ICEM)*, Lausanne, Switzerland, September 2016. New York, NY: IEEE; 2016, pp. 2704–2710.

[21] Schlicthing H. *Boundary-Layer Theory*, 7th edn. New York, NY: McGraw-Hill Series in Mechanical Engineering; 1979.

[22] Nakabayashi K., Yamada Y., and Kishimoto T. 'Viscous frictional torque in the flow between two concentric rotating rough cylinders'. *J Fluid Mech.* 1982;119:409–422.

[23] Vrancik J. E. Prediction of Windage Power Loss in Alternators – NASA Technical Note, *NASA TN D-4849*; 1968.

[24] Yamada Y. 'Torque resistance of a flow between rotating co-axial cylinders having axial flow'. *Jpn. Soc. Mech. Eng.* 1962;5(20):634–642.

[25] NEMA Standards, Publication No. MG7, Revision; 1993.

[26] Heiles F. 'Design and arrangement of cooling fins'. *Elecktroteckn. Maschinenbay.* 1952;69(14).

[27] Boglietti A., Cavagnino A., and Staton D. A. 'Thermal sensitivity analysis for TEFC induction motors'. *Second IET International Conference on Power Electronics, Machines and Drives*, Edinburgh, UK; 2004, pp. 160–165.

[28] Al-Timimy A., Giangrande P., Degano M., and Xu Z. 'Design and losses analysis of a high power density machine for flooded pump applications'. *IEEE Trans. Ind. Appl.* 2018;54(4):3260–3270.

[29] Malumbres J. A., Satrustegui M., Elosegui I., Ramos J. C., Martínez-Iturralde M. 'Analysis of relevant aspects of thermal and hydraulic modeling of electric machines. Application in an open self ventilated machine'. *Appl. Therm. Eng.* 2015;75:277–288.

[30] Knowlton E., Rice C. W., and Freiburghouse E. H. 'Hydrogen as a cooling medium for electrical machinery'. *Trans Am. Inst. Electric. Eng.* 1925;44:922–934.

[31] Snell D. S. 'The hydrogen-cooled turbine generator'. *Trans. Am. Inst. Electric. Eng.* 1940;59(1):35–50.

[32] Nategh S., Krings A., Wallmark O., and Leksell M. 'Evaluation of impregnation materials for thermal management of liquid-cooled electric machines'. *IEEE Trans. Ind. Electron.* 2014;61(11):5956–5965.

[33] Li H., Klontz K. W., Member S., Ferrell V. E., and Barber D. 'Thermal models and electrical machine performance improvement using encapsulation material'. *IEEE Trans. Ind. Appl.* 2017;53(2):1063–1069.

[34] LORD, Increase Power Density and Motor Lifetime with CoolThermTM Materials, LORD Corporation OD PB8198 (Rev.5 9/17); 2017.

[35] Madonna V., Walker A., Giangrande P., *et al.* 'Improved thermal management and analysis for stator end-windings of electrical machines'. *IEEE Trans. Ind. Electron.* 2019;66(7):5057–5069.

[36] La Rocca A., Xu Z., Arumugam P., *et al.* 'Thermal management of a high speed permanent magnet machines for an aeroengine'. *2016 XXII International Conference on Electrical Machines (ICEM)*, Lausanne, Switzerland, September 2016. New York, NY: IEEE; 2016, pp. 1–6.

[37] Sano S., Yashiro T., Takizawa K., and Mizutani T. 'Development of new motor for compact-class hybrid vehicles'. *World Electr. Veh. J.* 2016;8 (2):443–449.

[38] EL-Refaie A. M., Alexander J. P., Galioto S., *et al.* 'Advanced high-power-density interior permanent magnet motor for traction applications'. *IEEE Trans. Ind. Appl.* 2014;50(5):3235–3248.

[39] Leland J. E. and Pais M. R. 'Free jet impingement heat transfer of a high Prandtl number fluid under conditions of highly varying properties'. *ASME. J. Heat Transfer* 1999;121(3):592–597.

[40] Womac D. J., Ramadhyani S., and Incropera F. P. 'Correlating equations for impingement cooling of small heat sources with single circular liquid jets'. *ASME. J. Heat Transfer.* 1993;115(1):106–115.

[41] Davin T., Pellé J., Harmand S., and Yu R. 'Experimental study of oil cooling systems for electric motors'. *Appl. Therm. Eng.* 2014;75:1–13.

[42] Bennion K. and Moreno G. 'Convective heat transfer coefficients of automatic transmission fluid jets with implications for electric machine thermal management'. *ASME 2015 International Technical Conference and Exhibition on Packaging and Integration of Electronic and Photonic Microsystems (InterPACK2015)*, San Francisco, CA, USA, July 2015. ASME; 2015, pp. 1–9.

[43] Davin T., Pellé J., Harmand S., and Yu R. 'Motor cooling modeling: an inverse method for the identification of convection coefficients'. *J. Thermal Sci. Eng. Appl.* 2017;9(4).

[44] Assaad B., Mikati K., Tran T. V., and Negre E. 'Experimental study of oil cooled induction motor for hybrid and electric vehicles'. *International Conference on Electric Machines*, Alexandroupoli, Greece, September 2018. New York, NY: IEEE; 2018, pp. 1195–1200.

[45] DiGerlando A. and Vistoli I. 'Thermal networks of induction motors for steady state and transient operation analysis'. *Proceedings of the ICEM Conference*, Paris, France, 1994, pp. 452–457.

[46] Stokum G. 'Use of the results of the four-heat run method of induction motors for determining thermal resistance'. *Elektrotechnika* 1969;62 (6):219–232.

[47] Hamdi E. S. *Design of Small Electrical Machines*. Hoboken, NJ: Wiley; 1994.

[48] Wendt F. 'Turbulente Strömungen zwischen zwei rotierenden konaxialen Zylindern'. *Ingenieur-Archiv*. 1933;4:577–595.

[49] Bilgen E. and Boulos R. 'Functional dependence of torque coefficient of coaxial cylinders on gap width and Reynolds numbers'. *J. Fluid Eng.* 1973;95(1):122–126.

[50] Rahman K., Anwar M., Schulz S., *et al.* 'The Voltec 4ET50 electric drive system'. *SAE Int. J. Engines*. 2011;4(1):323–337.

Chapter 5

Advanced computational methods for modelling ventilation and heat transfer

The use of advanced computational methods: finite-element analysis (FEA) and computational fluid dynamics (CFD) will be considered in this chapter. FEA is useful to model conduction heat transfer, while CFD can additionally model fluid flow. Both methods rely on modelling the actual geometry of the full machine or machine component by discretizing the individual parts into smaller mesh elements, volumes or cells over which the heat transfer or flow variables are assumed constant. Triangular and quadrilateral mesh element shapes are typically used for 2D geometries, and tetrahedral, quadrilateral pyramid, triangular prism, and hexahedral shaped cells for 3D geometries.

The FEA method is often applied to individual components in the electrical machine where the heat transfer mode is by conduction, such as the slot and winding. Useful models can often be formed in 2D. 3D FEA models for a particular part can also be created but are less commonly used due to the additional time to set up and solve.

CFD is mainly of use to model the fluid flow and convection from surfaces in the machine and the models are nearly always 3D in nature. Models can be created for individual components of the machine or for the full machine, although this can be very complex to put together and solve. The model may also include conduction heat transfer.

5.1 Finite-element methods

As described in Chapter 2, conduction heat transfer can be modelled if the machine geometry and material selection are known. However, conduction heat transfer in certain regions is complex but important, for example, heat transfer in a slot. The slot contains several components (i.e. copper conductor, wire insulation, impregnation/potting material, slot liner, air voids) with significant differences in material thermal properties. Moreover, the conductors are typically distributed randomly within the slot as illustrated in Figure 4.5. Therefore, even with a superficial knowledge of the geometrical layout and material properties, it is often not possible to obtain an accurate prediction of the conduction heat transfer in a slot using a lumped-parameter thermal network (LPTN).

As a solution, FEA can play an important role in making accurate predictions of conduction heat transfer for complex geometric shapes. Moreover, FEA is a standard tool for detailed and accurate loss determination. Consequently, when coupling electromagnetic and thermal FEA models, accurate thermal solutions can be obtained. However, it is important to note that FEA is not capable of solving the interface and convection problems. The issue of interfaces has been discussed in Section 4.1.6 and several empirical studies have been done to address the factors that affect the contact resistance between machine components. For convection heat transfer, in a FEA model, the convective boundaries need to be pre-defined from either published convection heat transfer correlations or CFD simulation results.

To perform thermal analysis of a full machine using FEA method can be time consuming. Due to these limitations of FEA, it is recommended to use the FEA method to calibrate a lumped circuit model, e.g. thermal resistances used in the thermal circuit to give the same temperature rise. Additionally, the FEA method can be a useful tool to understand the location of the maximum hot spot in the winding and this can provide insights into where to apply cooling, e.g. direct slot cooling.

5.1.1   Stranded random wound winding

Stranded random wound windings are one of the most common winding types used in electrical machines and these are often known as distributed windings. For distributed windings, the detail of where the conductors are placed in a slot is not known exactly as they are randomly distributed and spread over the slot area. Before the conductors or windings are inserted into the slots, the slots are insulated with insulation paper, known as the slot liner. Then the space between the conductors is filled with impregnation material, e.g. epoxy resin, varnish, etc. By estimating the placement of the copper conductors in a slot, a 2D thermal FEA model can be created as depicted in Figure 5.1 with discrete regions for each conductor. The mesh in the slot is denser because of the thermal gradient is high and hence more computational elements are required.

To obtain a steady-state thermal solution, a fixed temperature boundary condition can be applied to the outer diameter of the stator lamination. For heat sources, the winding loss and stator iron loss can be applied to the regions where they are generated. Since only one stator slot segment is modelled, the periodic planes of the segment can be treated as adiabatic. This assumes that each segment of the stator is subjected to the same amount of loss and cooling. In Figure 5.2, the 2D thermal FEA solution for a given thermal boundary condition and loss is obtained. The winding hot spot is located at the centre of the slot because the insulation materials have low thermal conductivity which restricts the heat conduction to the stator lamination.

As described in Section 4.1.1, the winding heat transfer can be calculated analytically using a lumped circuit model with anisotropic thermal conductivities. The values of the equivalent thermal conductivities are different depending on the direction of heat transfer. For heat transfer across a coil, the thermal conductivities in the radial and tangential direction can be assumed to be the same for a distributed winding.

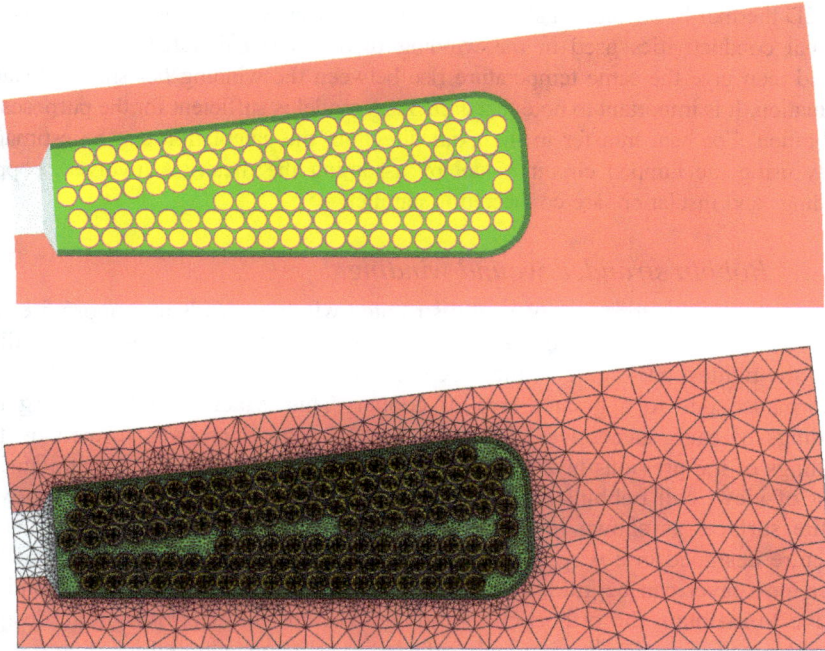

Figure 5.1   2D thermal FEA model geometry and mesh for a stator slot segment

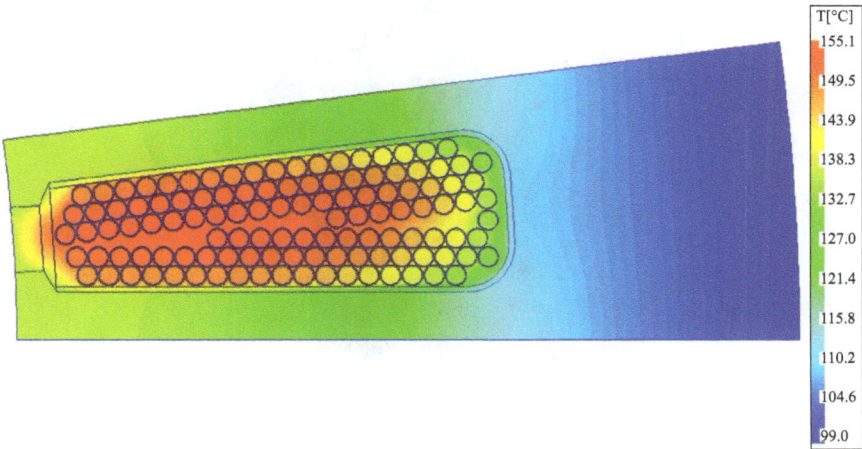

Figure 5.2   2D thermal FEA solution for a stator slot with distributed stranded winding

The 2D thermal FEA solution allows the machine designer to calibrate the equivalent thermal conductivities used in the winding model. The calibrated winding model should then give the same temperature rise between the winding hot spot and stator lamination. It is important to note that a 2D FEA model is sufficient for the purposes of calibration. For heat transfer in the axial direction, the calculation can be estimated easily using the lumped circuit model by assuming the materials involved, copper windings and insulation, are connected in parallel.

5.1.2   Bobbin stranded wound winding

The random conductor placement of distributed windings leads to a poor wire slot fill factor. In order to achieve a more optimal wire slot fill, an orthocyclic winding gives the tightest packaging of the wire cross-section [1] and thus a better thermal performance due to a greater area of copper. This is achieved by placing the windings of each successive layer into the "V"-shaped grooves formed by the windings of the previous layer, as illustrated in Figure 5.3. From a manufacturing point of view, coil bobbins are installed in the stator lamination with a grooved profile surface, as shown in Figure 5.3. The coil bobbin is typically a plastic structure designed in a way to ease the winding process, so that the wires of the inner most layer can be guided across the profiled surface into the corresponding groove to maintain the orthocyclic winding structure. In electromagnetic design, the orthocyclic winding is commonly referred to as a concentrated winding type or

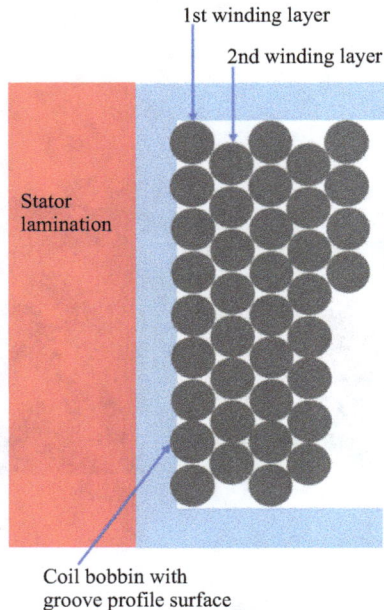

Figure 5.3   Schematic diagram for orthocyclic winding

tooth wound winding type. For both winding types, the wires are forced toward the stator tooth as depicted in Figure 5.3. As the stator lamination has typically a much lower temperature than the coils, this leads to better heat transfer due to the reduction in thermal path length. The difference between a distributed winding and a concentrated winding can be visualized in Figure 5.4.

For thermal modelling, if the details of where conductors are placed within a slot are known it is easier to make an accurate thermal model than in the case of a distributed winding. When coil bobbins are available, some designers take advantage of this to replace the slot liner by extending coil bobbin structure into the stator slot to save production processing and material costs. However, there is a drawback in that the bobbin in the stator active section is usually much thicker than insulation paper (slot liner) due to the limitations of the plastic injection moulding technology. In Figure 5.5, the 2D thermal FEA solution for a stator using a tooth wound winding is presented. Due to the thickness of bobbin, there is a significant temperature rise between the stator and the coil. The thermal FEA solution can be used to calibrate the winding model.

5.1.3   Hairpin winding

Due to increasing demands of torque and power density in electric traction motors, hairpin windings are becoming popular and they have been widely employed in electric motors for battery and hybrid electric vehicles such as Chevrolet Spark [2], Chevrolet Volt [3], Chevrolet Bolt [4], and Toyota Prius [5]. Compared to distributed stranded windings, hairpin windings give the following advantages:

(i)   High slot fill
(ii)   Reduction in DC electrical resistance
(iii)   Improved thermal performance due to increased copper area

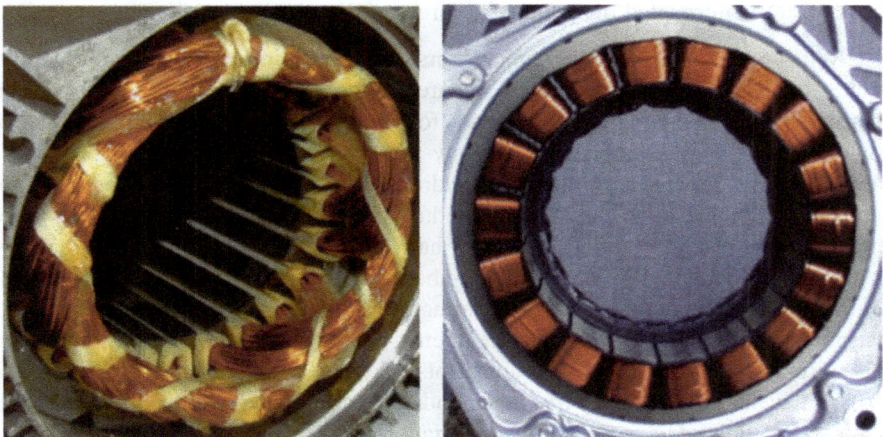

Figure 5.4   Distributed winding and concentrated winding [19]

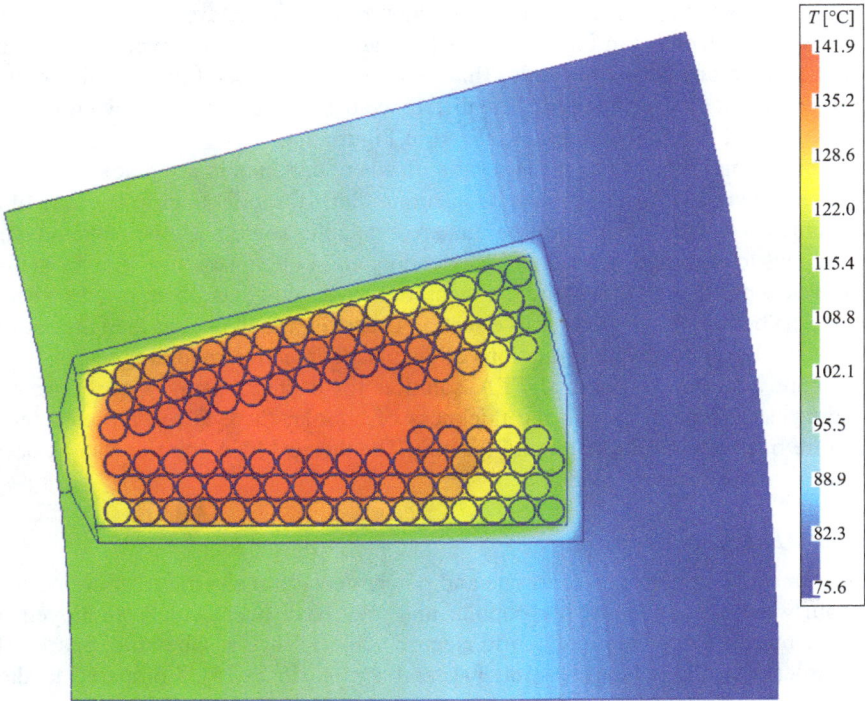

Figure 5.5   2D thermal FEA solution for a stator slot with tooth wound winding

(iv)   Shorter end-turn length
(v)   Improved active cooling performance in the end-turn due to the gaps formed between the hairpins leading to a larger surface area
(vi)   A fully automated manufacturing process

With rectangular-shaped copper bars, hairpin windings can significantly reduce the DC electrical resistance compared to the round section wires adopted in distributed stranded windings. This leads to an increase in motor efficiency due to lower copper losses, especially at low- to medium-speed operating ranges. However, the drawback of hairpin windings is extra AC winding loss due to induced eddy currents in the conductors during high speed operation, i.e. the skin effect and the proximity effect [6,7]. For hairpin windings, a parallel slot topology is used which better suits the rectangular-shaped conductors, rather than the parallel tooth topology used for stranded windings. For bar wound stator construction, hairpins are made in a "U"-shaped geometry, as shown in Figure 5.6(a), and then the hairpins are inserted into the stator slots. Then they are twisted (see Figure 5.6 (b)) and welded together at the other end to form a wave-winding pattern in a completely automated process. It is important to note that this construction leads to an asymmetric end winding geometry and thus a different copper loss distribution between the drive and non-drive end.

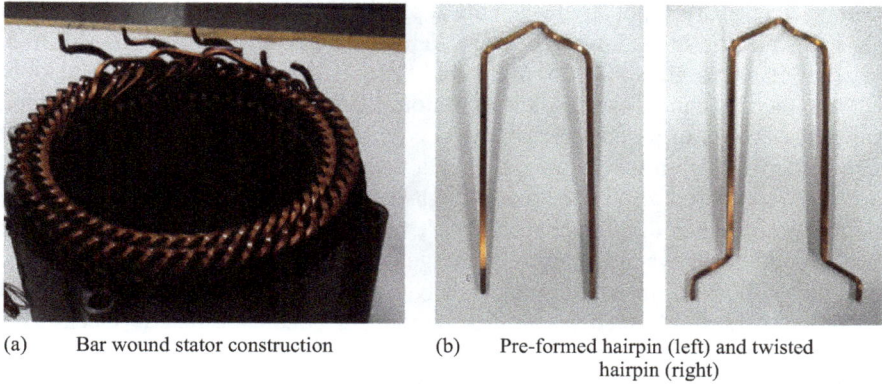

(a)   Bar wound stator construction   (b)   Pre-formed hairpin (left) and twisted
                                                      hairpin (right)

*Figure 5.6   Hairpin winding of Chevrolet Spark battery electric vehicle
propulsion electric motor [2]*

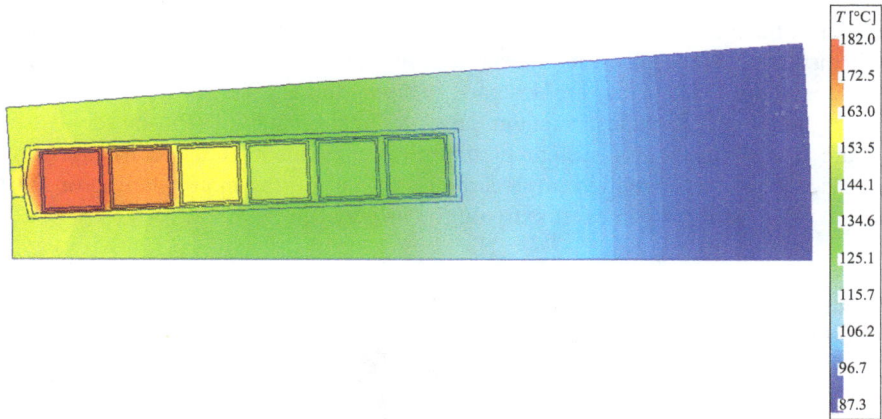

Figure 5.7   2D thermal FEA solution for a stator slot with hairpin winding

Figure 5.7 shows the 2D thermal FEA solution for a stator with six rectangular conductors in a slot. Compared to a stranded winding, a hairpin winding can utilize the space in the slot more effectively. By coupling with the FEA loss calculation, the thermal FEA model can provide an accurate prediction of winding heat transfer in the slot. As shown in Figure 5.7, the temperature rise of the conductors near the slot opening is much higher due to the AC winding loss. Although the equivalent thermal network can be built easily as the geometry of hairpin winding is relatively simple, the winding model can be also calibrated by the thermal FEA solution.

5.1.4   Pre-formed wound winding

For high-voltage machines such as turbo generators for the power industry, a special high-voltage AC insulation system is required. In fact, all electric insulators become electrically conductive when the applied voltage is sufficiently large which leads to dielectric breakdown. To avoid dielectric breakdown, materials with high resistivity such as glass, Teflon, Mica, and Kapton are usually used as insulation. High-voltage machines can breakdown before the breakdown voltage is exceeded due to the effects of partial discharge and corona discharge. Partial discharge can occur at an insulation weak spot or when a defect in the insulation during manu-facturing is exposed to an excessively high-voltage stress and a partial discharge occurs. With corona discharge, special care needs to be taken to limit the electric field strength, even when the applied voltage is not high enough to cause dielectric breakdown, because in a high-voltage environment the electric field can cause ionization in a fluid, such as air, surrounding a conductor that is electrically charged.

For high-voltage machines, the windings are pre-formed so that special high-voltage insulation systems can be used. The pre-formed windings normally use rectangular wire and a parallel slot topology. Figure 5.8 shows pre-formed wind-ings inside a slot which contains two coils in a double layered winding. One coil at the slot bottom is separated from the other coil at the slot opening by a phase insulator. Each coil is formed by 8 rectangular wires and they are protected by thick coil insulation. The coil insulation is usually made of multiple insulation layers. Wedges at the slot opening are designed in a way to keep the coils in place. Also, for high-voltage machines, vacuum pressure impregnation (VPI) is adopted to ensure a rigid insulation system to minimize movement at the coil-lamination interface when there are high surge currents, such as during motor starting or short circuit faults as these can exert extreme forces between the coils.

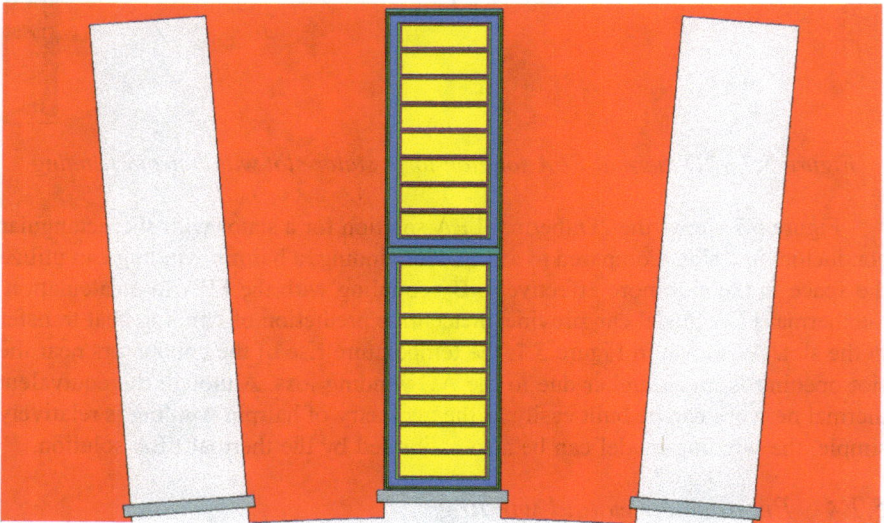

Figure 5.8   Pre-formed winding in a stator slot

Figure 5.9 *2D thermal FEA solution for a stator slot with pre-formed wound winding*

The complete insulation system for high-voltage machines consists of different insulation materials, such as wire insulation, coil insulation, phase separators, slot liners, impregnation, coil spacers, etc. They all have different thickness, thermal properties and even material temperature limits. To avoid insulation breakdown, designers need to make accurate predictions of the machine hot spot temperature and its location. As shown in Figure 5.9, an FEA model can be useful to calibrate the winding model for high-voltage machines.

5.1.5   Litz wires

For high-frequency operations, electrical machines are subjected to eddy current effects in the winding [6,7]. The skin effect and proximity effect lead to an AC winding loss which is an additional loss in the conductors, especially conductors with a large cross-sectional area, such as hairpin windings. This extra winding loss affects not only the machine efficiency but also the machine thermal performance. Due to the nature of the AC winding loss, it is mostly concentrated at the region near slot opening. This leads to high thermal gradients and the winding hot spot occurs in this region.

To mitigate the AC winding loss, Litz wire is used which can significantly reduce the eddy currents induced in the conductors [8]. Litz wire is constructed from a group of individually insulated strands of fine wires which are twisted together in a bundle. For a given winding area, the amount of copper in a Litz wire winding will potentially be less than it could be with a solid wire winding, due to the insulation of each single wire strand. This leads to higher DC (Ohmic) electrical resistance than that of a solid wire. Moreover, the actual length of the Litz wire is longer due to the effect of twisting in the bundles. Hence, the twist of Litz wire can further increase the DC electric resistance. To reduce the drawback of this increase in DC electrical resistance, Litz wire is usually compressed into a rectangular cross-sectional profile for tighter packing. Typically, the Litz bundle is insulated with an outer insulation layer to maintain the compressed profile and also to

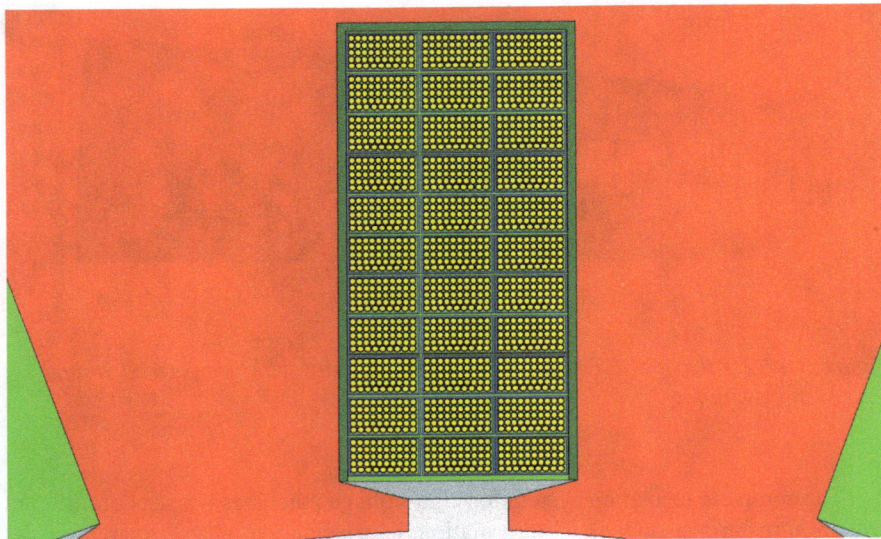

Figure 5.10   Litz wire in a slot

prevent dielectric breakdown and for environmental protection, i.e. type 8 Litz wire offered by New England Wire [9].

Figure 5.10 illustrates the use of rectangular cross section Litz wire in a stator slot with a parallel slot topology. Each Litz bundle comprises of a number of fine strands. This causes some difficulties in creating a representative FEA model in attempting to model every single strand individually. In an alternative solution, a FEA model as shown in Figure 5.11 can be created to give accuracy at the bundle level rather than strand level. To examine the conduction heat transfer at strand level, a separate FEA model is suggested to calibrate the values of equivalent thermal conductivity used in FEA model at bundle level.

To account for the complex construction of Litz wires, experimental testing is required to validate the equivalent thermal conductivities as described in Section 6.2. When compared to the FEA method, experimental methods can address the influence of Litz wire construction on thermal conductivity more effectively.

5.2   CFD

5.2.1   Introduction to CFD

In most electrical machines, not only is conduction heat transfer important, but heat is also dissipated by convection to a cooling fluid. This could be by convection from the outside of the frame or through the use of a liquid cooling system. Analysis of convection heat transfer has been described in Chapter 2 and this

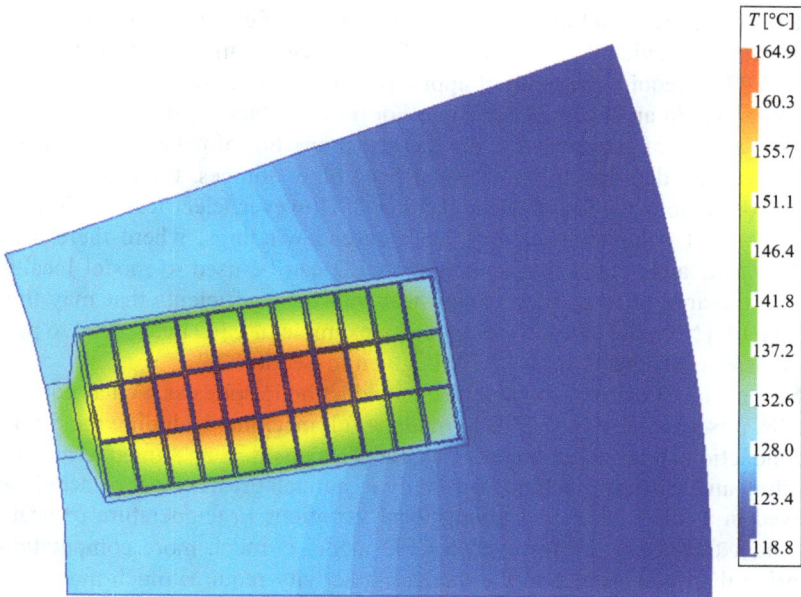

Figure 5.11   *2D thermal FEA solution with bundle level accuracy for a stator slot*
with Litz wire

involves an understanding of the fluid flow, described in Chapter 3. For straight-forward situations, such as natural convection from the outside of simple shapes and forced convection in straight ducts, fluid flow and convection heat transfer can be calculated analytically as described in Chapters 2 and 3. In other situations the fluid flow and associated convection heat transfer from solid surfaces requires a more complex treatment. The governing equations that describe the physics of flow and heat transfer in fluids are the Navier–Stokes equations as described in Chapter 3. These equations can only be solved analytically for very simple situations such as laminar flow over a flat plate or in a long circular duct and these situations rarely occur in practice. CFD is the method by which the Navier–Stokes equations can be solved numerically and applied to the practical geometries and fluid flow arrangements that are found in engineering. CFD allows the fluid flow and associated convective heat transfer to be modelled by dividing up a region of fluid flow into discrete volumes, or cells, in which the flow variables (e.g. pressure, velocity, temperature) can be assumed to be uniform. The Navier–Stokes differential equations can then be written in a simplified form as a set of linear algebraic equations that can be solved to describe the relationship between the flow variables in adjacent cells. The most popular numerical technique for CFD is the finite volume method and this is fully described in Ref. [10]. This is the method adopted by the CFD software tools that are commonly used in industry, for example, in the ANSYS suite of CFD tools: Fluent [11] and CFX [12].

CFD can be used in three ways in the modelling of electrical machines. First, it can be used to provide data on heat transfer coefficients and fluid flow for use in a LPTN. LPTNs require the input of appropriate thermal resistances to represent the heat flows within an electrical machine. For heat conduction, these can be derived from knowledge of geometry and the thermal properties of materials, however, for heat convection data has to be obtained from other sources, for example, experimental correlations as described in Chapter 2. However, electrical machines contain regions of complex geometry, such as end windings, where there are few correlations available. In these situations, CFD can be used to model locally the flows in the area of interest to obtain heat transfer coefficients that may then be used in a LPTN model of a full machine. An example of CFD being used to provide heat transfer coefficients for a LPTN is given in Section 7.7.

Second, as CFD can also model heat conduction in solids as well as convection in fluids, it is possible to use CFD to model a full machine, including the fluid flow and conduction and convection heat transfer. This could be an attractive method for modelling an electrical machine, especially as a much greater level of detail can be achieved in a full CFD model giving local variations in temperature or heat flux throughout the machine. However, a CFD model is much more computationally expensive than LPTNs and so the modelling activity requires much more computational resource and takes more time for both model setup and computation.

Finally, CFD can also be used to model the windage losses in electrical machines. This is done through integrating the torque generated on the solid rotating surfaces in a machine arising from the pressure and frictional shear stresses within the fluid. Not only can the windage torque on the rotor be determined, but torque required to drive fans can also be modelled. It is thus possible to design optimised cooling fans for electrical machines as described in Ref. [13].

CFD modelling has been used to provide data for LPTN models since the mid-1990s and it is now used widely in this way in the design of electrical machines. With advances in computational power, modelling a full electrical machine in CFD is now becoming possible on high-performance desktop computers. However, it is expensive to use in terms of labour and computational time and it is likely to be some years before it replaces the use of LPTNs for general purpose electrical machine modelling.

In the next two sections, the application of CFD to electrical machines will be described. First, in Section 5.2.2, the general procedure for using CFD will be described. Then, in Section 5.2.3, specific issues in using CFD for modelling electrical machines will be described and examples of its application given.

5.2.2   Using CFD

There are four main stages in undertaking a CFD modelling activity:

- Defining and creating the geometry
- Generating the computational mesh
- Computing the solution
- Post-processing and interpreting the results

It should be mentioned that CFD modelling is usually an iterative process and as the stages proceed issues arise that require making changes in earlier stages. It should not be expected that the stages can be followed in straight succession except in very simple cases.

5.2.2.1　Defining the geometry

The geometry required for a CFD model can usually be imported from appropriate CAE software and so does not normally have to be created from scratch. However, the information in a CAE model is typically much more detailed than is required and some simplification will be needed. For instance, when modelling the flow over the outside of a fan cooled machine, features in the geometry that do not have a significant influence on the flow can be omitted for example bolt heads and all the interior surfaces and small recesses that that do not affect the flow may be omitted. In simplifying the geometry of a machine some knowledge of fluid mechanics is desirable to be able to make appropriate simplifications that will not significantly affect the fluid flow. Rough surfaces on the outside of laminations may be modelled as a general surface roughness rather than the edge of each lamination.

It may also be possible not to model a full machine but assume axes of symmetry to reduce the size of the model required. If the region of interest is axi-symmetric with no parameter variations around the circumference then a 2-dimensional model may be appropriate. If there are regularly occurring features such as stator windings or pole winding, then only a sector of a machine may need modelling and periodic boundary conditions can be used to represent the rest of the machine. For instance, if a fan-cooled motor has a fan with 13 blades, it may be easier to simplify it to 12 blades and then model a 30° sector rather than full 360°. Or a 90° sector could be modelled for a 4 pole machine.

If part of a machine is being modelled, the model will need to extend as far as areas where the boundary conditions are known. For instance, up to a solid wall or up to a duct inlet where the flow rate is known. When modelling external flows over machines, it will normally be required for the model to extend sufficiently far upstream and downstream from duct inlets and outlets for the velocities and pressures to be uniform over a boundary.

If conduction in solid regions of a machine is also to be modelled, then simplifications can also be made in these regions for instance to remove interfaces where there is little heat flux or modelling thin layers of insulation as "thin surfaces" with a given thermal resistance.

5.2.2.2　Creating the mesh

In a CFD model, the variables are calculated discretely in each cell by solving the governing equations. To be able to obtain an accurate representation of the flow in a region where there are significant variations in velocity or temperature, the mesh has to be suitably fine. However, the fineness must be balanced against the computational effort as the time needed to solve a problem will generally be proportional to the number of cells. The usual process is to make the mesh sufficiently fine that it does not influence the CFD solution and several mesh refinements may

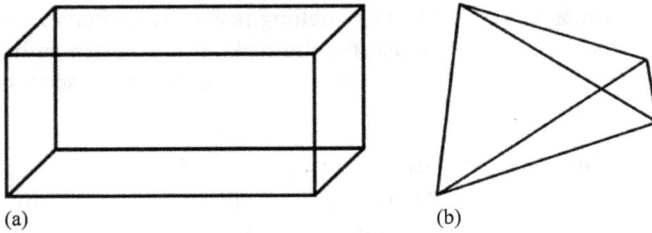

(a)   (b)

Figure 5.12   Representation of (a) hexahedral and (b) tetrahedral shaped cells

be necessary to find a result that is mesh independent. Sometimes it can be appropriate to use adaptive mesh refinement within a CFD problem where the mesh can be refined automatically in regions where there are high gradients in velocity or pressure.

There are generally two shapes of cell that may be used in a mesh: four-sided tetrahedrons (pyramid shaped) and six-sided hexahedrons (brick shaped), as shown in Figure 5.12. Tetrahedral cells are most suited to modelling irregularly shaped regions as they can easily be arranged to confirm to the geometry. They will usually be appropriate for meshing the end regions of a machine. The mesh density can easily be increased around particular features in the flow, such as the end windings, coils and fluid duct inlets and outlets. Hexahedral cells are appropriate for more uniform areas of geometry such as ducts and the air gap. These cells can be stretched so that they are longer in the direction where gradients in the flow are small and shorter in regions with strong gradients. An example would be the air gap of a machine or in a duct where it would be usual to have at least 10 or 20 cells across the air gap or a duct to be able to resolve the flow profile. Cells can be much longer along a duct or axially along the air gap where the gradients are smaller. Hexahedral cells do not have to be cuboid in shape but can have unequal angles at the vertices.

Usually, in a CFD model, the domain must be split up into a number of blocks and different types and density of cell can be used. At the interface between the blocks, the meshes on either side may match exactly in size and shape, known as a conformal mesh, or they may be of different shapes and sizes, known as a non-conformal mesh. A non-conformal mesh requires interpolation across the interface and this may introduce inaccuracies if the boundary is in a region of a strong gradient. Generally hexahedral cells will be required to achieve a conformal mesh across a block boundary. Conformal and non-conformal meshing across an interface is illustrated in Figure 5.13.

There are very strong gradients in velocity near to a wall as a fluid is always assumed to be stationary at a wall. This is a fundamental principle of fluid flow known as the non-slip boundary condition. It would be very demanding to generate a mesh capable of resolving the flow profile at the boundary, particularly for turbulent flows, and CFD codes employ wall functions to deal with this boundary condition. For these wall functions to be implemented accurately, it is necessary for

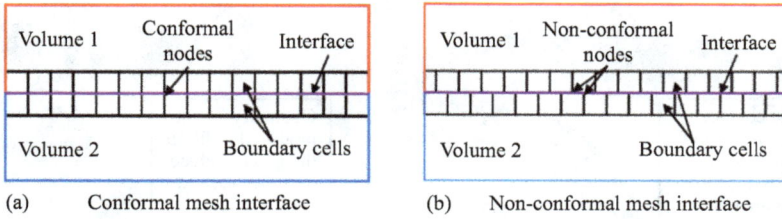

(a) Conformal mesh interface   (b) Non-conformal mesh interface

Figure 5.13   Illustration of (a) conformal and (b) non-conformal meshing at an interface (with permission from Ref. [20])

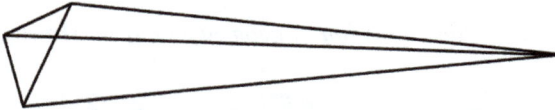

Figure 5.14   A tetrahedral cell with substantial skewness

the cells at the wall boundaries to be of a certain size usually denoted by a non-dimensional distance, usually denoted by y^+, that relates to the fluid flow conditions.

Cells do not need to have equal angles at all the vertices and they may be skewed to suit the geometry. However, if the variation between the angles is large and the cell is significantly skewed this can result in errors. A tetrahedral cell with a substantial skewness is shown in Figure 5.14. Usually a CFD code will identify cells where the skewness is not acceptable and either the mesh can be refined or sometimes, it is appropriate to adjust the geometry to avoid this problem. A skewness problem that may occur in electrical machines is where a circular shape is in contact with a flat surface. At the interface, the angle at the vertex of the cell tends toward zero and leads to errors. This can be avoided simply by adding a small fillet radius at such points of contact. This will not affect the flow but will prevent what can otherwise be troublesome errors. This is illustrated in Figure 5.15.

Heat conduction in solid regions can be modelled by CFD and in these regions a courser mesh can often be used. Figure 5.16 shows two examples of a mesh in which heat conduction in the rotor and stator is modelled as well as air flow. Note that a much finer mesh is used in the regions where air flow is modelled. In general, a hexahedral mesh is used for the solid regions and a tetrahedral mesh for the regions where there is air flow.

Once the mesh has been created the final stage in setting up, the model is to define the boundary conditions for the CFD calculations. These will typically involve the following:

• Defining surfaces which are solid walls. These will be stationary for the stator and other non-rotating parts of a machine and moving walls for the rotating parts. The modelling of rotation will be dealt with in more detail in Section 5.2.3.

At the contact between a round and flat shape there will be a cell with a vertex angle tending to zero, causing skewness problems.

Geometry can be modified to fill in the gap to reduce need for skewed cells at point of contact.

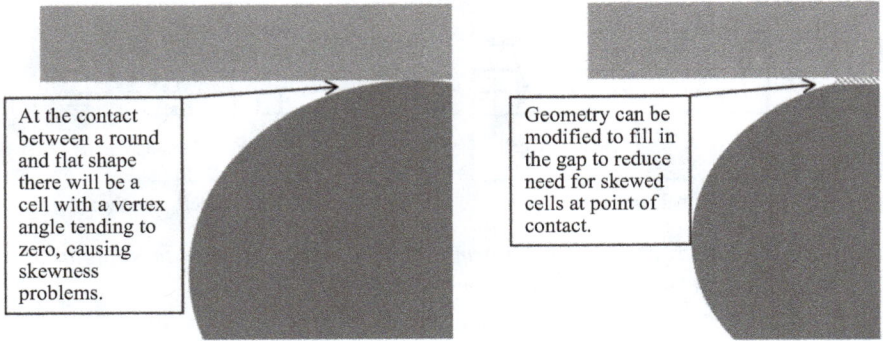

Figure 5.15   Simplification of geometry to avoid skewed cells

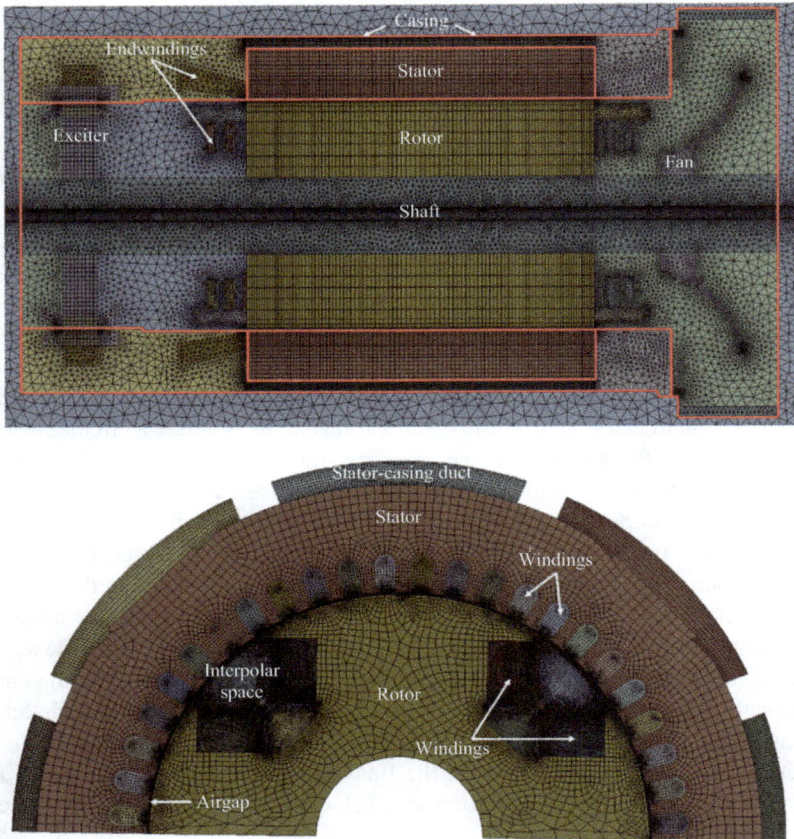

Casing

Endwindings

Stator

Exciter

Rotor

Fan

Shaft

Stator-casing duct

Stator

Windings

Interpolar space

Rotor

Windings

Airgap

Figure 5.16   Example of CFD mesh for an electrical machine (with permission from Ref. [20])

- Defining the conditions at flow inlets and outlets. A flow inlet can be defined in terms of a pressure or velocity profile and outlets can also be determined in terms of a pressure, if the inlet has been defined in this way or a mass flow if the inlet flow is prescribed.
- Defining boundaries that are planes of symmetry, across which there is no flow or periodic boundaries where the flow leaving one boundary is matched by a flow entering on the corresponding boundary. Periodic boundary conditions are used whenever a sector of a machine is being modelled and the periodic boundaries are on the radial planes that cut the flow domain. Periodic boundaries occur in pairs and there can be an outflow or inflow at one boundary but it is matched by a corresponding inflow or outflow at the opposite boundary. This allows for swirling flows to be modelled when only a sector of a machine is being modelled. At a symmetry boundary, no mass flow or heat flow can occur across the boundary and so velocities and temperature gradients are zero normal to the boundary. Symmetry boundaries are used much less in electrical machines as there is rarely a plane of geometrical symmetry across which there is no flow. An example where a symmetry plane might be used could be an air cooled machine where air enters at each end and leaves in the middle. A plane of symmetry perpendicular to the axis in the middle of the machine may then be appropriate.

5.2.2.3 Computing the solution

Once a CFD geometry, mesh and boundary conditions have been defined the problem can be solved. The from the mesh and boundary conditions the CFD code will numerically solve the governing Navier–Stokes equations describing the physics of the fluid flow including:

- Conservation of mass
- Conservation of momentum
- Conservation of energy

This is done within the CFD software by deriving linear algebraic equations for each cell that approximate to the governing differential equations to allow the physical variables (velocity, pressure, temperature, etc.) in each cell to be calculated. The discretization of the governing equations can be done to a first order approximation in the simplest cases, but higher order, more accurate, approximations can be used for better accuracy, although this is often at the expenses of robustness and computational effort. The equations for each cell are solved iteratively to converge to the best solution that gives the best fit of the variables to the governing equations. The CFD user will be able to control a range of parameters in the solver to control the solving process and guidance on the use of these will be available from the CFD software vendor. Some guidance for common issues related to electrical machines is as follows:

- Turbulence cannot be dealt with in the governing equations as the scale of the chaotic time-dependent velocity fluctuations, the turbulent eddies, is usually very much smaller than the computational mesh and so the turbulence cannot be resolved. This is most commonly overcome by averaging the turbulent fluctuations and modelling the effects of turbulence on the flow in terms of the effective increase in viscosity and energy dissipation that it gives. This is done in the so-called Reynolds Averaged Navier–Stokes equation CFD codes (RANS CFD) and a range of turbulence models will be available to the user. The most widely used CFD codes such as ANSYS Fluent and CFX use RANS solver. The most common turbulence model is the two equation k–ε model which has been found to give good results for most flows found in electrical machines. The alternative k–ω turbulence model has also been found to give good results particularly for internal flows bounded by walls.

- In most air cooled machines, the Reynolds (Re) in the air flows will generally be in the turbulent region (Re > 3,000 in ducts) and so modelling of turbulence will be important.

- Where very high speeds occur in a flow, then the compressibility of the fluid may need to be taken into account. However, in general in electrical machines velocities are generally low and Mach numbers (velocity relative to the local speed of sound) are generally lower than 0.3, the value above which compressibility effects generally need to be taken into account. For electrical machines using air cooling the speed of sound will be about 350 m/s and so compressibility effects will only be significant if there are velocities (e.g. peripheral rotor speeds) in excess of 100 m/s within the machine.

- Where fluid flow over a solid surface occurs, boundary layers within the flow build up as shown in Figure 3.1, a very fine mesh would be needed to represent the velocity profile within a boundary layer and this is avoided in CFD codes by using *wall functions* to describe the effects of the boundary layer on the fluid wall shear stress and wall heat flux.

- CFD solvers use an iterative approach in calculation procedure. This involves starting from some initial assumptions for the values of the variables to be determined and then solving the equations repeatedly to find values for the variables in each cell that give the best fit to the governing equations in each cell. Residuals are calculated which represent the difference between the value of a variable within a cell and the value calculated from the governing equations based on the values of the variables in the neighbouring cells. A converged solution is determined to have been found when sufficient iterations have been completed to reduce the average residuals to a suitably low number. Typically for velocity, a residual of 10^{-3} is considered acceptable. This means that the velocity in each cell is on average within one part in a thousand or 0.1% of the values determined from neighbouring cells. This is generally acceptable for engineering calculations. In iterating to a converged solution, the speed of progress may be controlled by the degree to which the variables in each cell are corrected in each successive iteration, through choice of relaxation factors. Larger values of relaxation factors may hasten convergence, but in some cases, may cause a solution to diverge. Small relaxation factors mean that more iterations are required, but the convergence process will generally be more stable. Where temperatures are being calculated as well as flows, the flow calculations usually take longer to converge and it is often advantageous to calculate the flow on its own before initiating the energy equation to calculate temperature.

5.2.2.4   Post-processing and interpreting the results

CFD software usually has a large number of ways in which the results of an analysis may be presented for interpretation. Having completed a CFD analysis time should be set aside to investigate the most appropriate ways of visualising the results. Some of the most common ways of displaying the results are:

- Temperature is usually displayed using contour plots drawn through various planes that cut through the geometry under investigation. These contours may show temperature in the fluid and or solid regions. Alternatively, temperature contours may be shown on solid surfaces such as the surface of the rotor or casing.
- Heat flows are often best represented as contours of heat flux on solid surfaces from which convective heat transfer takes place.
- Heat transfer coefficients can be problematic to display in CFD software, particularly on surfaces within enclosures. Heat transfer coefficients do not represent a physical quantity. They are derived from a heat flux divided by a temperature difference and the temperature difference is that between the solid surface and some other reference temperature. Usually CFD software will require a reference temperature to be defined at a particular location. But heat transfer coefficients defined from experiment may be expressed in terms of average temperatures, for instance heat transfer from an end winding to the end space where an average temperature of the fluid throughout the end space is used. In ducts, the heat transfer coefficient is usually expressed in terms of the local temperature difference between the surface and the bulk fluid at a particular location in a duct. In taking data from a CFD analysis to provide heat transfer coefficients for LPTN models, it may be necessary to process the CFD results manually to obtain heat transfer coefficients that can be properly used in an LPTN model. For instance, for an end winding, the average bulk end region fluid and end winding surface temperatures can be output from CFD and then an overall heat transfer coefficient can be calculated by dividing the total heat flux by the mean temperature difference.
- Ways in which the fluid flow data can be displayed could be as velocity vectors and pressure contours. Mean velocities can be determined for ducts or openings to give a mass flow rate.
- Windage losses comprise two components: pressure drag and frictional drag. It is usually the pressure drag that dominates and this can be determined by summing the torque generated by the pressure forces acting in the tangential direction on a rotating surface. Most CFD software will have the capability to evaluate this. Frictional drag is of most significance on cylindrical surfaces and this is evaluated by summing the torque generated by the shear stresses on a surface.

5.2.3   Specific issues concerning the application of CFD to electrical machines

5.2.3.1   Complex geometries

Electrical machines can be difficult to represent in CFD as they have complex geometrical features some of which must be modelled in detail as they have important influences on thermal management. Other features however can be neglected.

In the air gap, there is significant heat transfer by convection between the rotor and stator and, in some smaller machines, the main cooling mechanism. The air gap is small relative to other geometrical features in a machine, but a fine CFD mesh is required to accurately model the flow and heat transfer. This is best done using regular hexahedral (brick shaped) cells and typically in the region of 20 cells would be needed in the radial dimension across the gap. In the circumferential directions, a fine mesh is also required to resolve the details of the flow occurring in the slots. However, a much coarser mesh can usually be used in the axial direction as the velocity and temperature gradients are much smaller in that direction. Figure 5.16 shows a typical mesh for an electrical machine. An example of the mesh in the air gap of an electrical machine is shown in Figures 5.17 and 5.18.

Optimal mesh density: A finer mesh will generally give a more accurate solution and this is needed in regions where there are high gradients in velocity or another variable. However, a high mesh density will also have more cells and require more computational resource and is likely to take long to solve. The normal procedure would be to use as coarse a mesh as possible and so undertake an exercise to progressively increase the mesh density until it no longer affects the solution. The solution is then said to be mesh independent. Most CFD codes now have the ability to automatically adapt the mesh density in regions of high-velocity gradient so that a fine mesh is not required for all of the domain being modelled.

Stator end windings have complex geometries and limit the thermal performance of a larger machines, so it is important to model the convective heat transfer

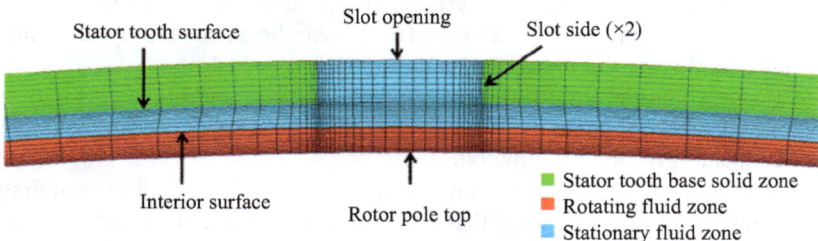

Figure 5.17   Mesh in the air gap of an electrical machine. The boundary between the rotating reference frame for the rotor and stationary reference frame for the stator is in the middle of the air gap. Note the higher mesh density near to solid surfaces (with permission from Ref. [15])

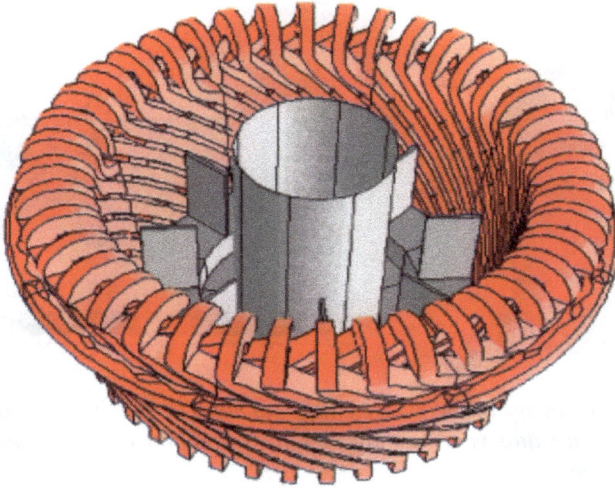

Figure 5.18   CFD model of the end winding in a high-voltage machine

with some accuracy. Generally, end windings fall into one of the two categories. In low-voltage machines, the stator coils are generally wound from wire and the end winding has a somewhat random structure in which the wire bundles in the coils are often manually bound to form the overall, often irregular, structure. In low-voltage machines, each end winding may be unique in the detail, due to the random way in which the coils are wound. Simple representations in CFD have included representing the end winding as a simple toroid in shape where the toroid has the same shape as the end winding envelope. These end windings are generally low in porosity but some porosity in the region where the coils emerge from the stator core can be represented. Although these simple shapes only give an approximation, they can nevertheless give an improvement on the basic experimental correlations that have hitherto been available. More detailed representations of low-voltage end windings are described in the case study below.

High-voltage machines tend to have coils that are formed from copper strip and although complex in shape these coils have a more defined shape and the overall end winding structure is more uniform. These end winding structures can be modelled in detail and an example is shown in Figure 5.18. The detail of the flow and heat transfer can be properly represented and conduction within the coils as well as convection to the air modelled in CFD.

However, a fine mesh is required to fully capture the details of the geometry and it is usual to use tetrahedral cells for these irregular shapes. Typically the mesh density will require 10–20 cells in between the coils so that the air flow and convective heat transfer can be accurately modelled. A number of investigations modelling high-voltage end windings have been reported and these have been used successfully to provide data to improve thermal design of end windings [21].

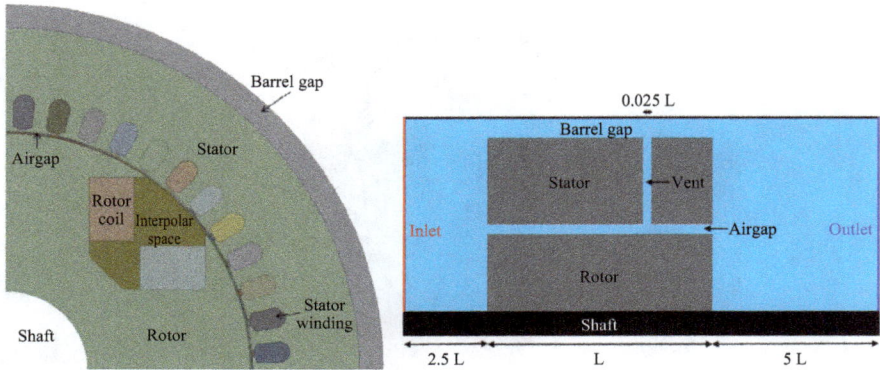

Figure 5.19   Geometry of the synchronous machine in which the optimal location of a stator vent was investigated to minimize stator winding temperature [18]

Other detailed geometrical features of a machine can often be neglected as they have little influence on the flow or heat transfer in a machine. Examples would include bolt heads, small recesses, or ribbed features.

5.2.3.2 Dealing with rotation and time-dependent flows

The rotation of an electrical machine normally generates important flows within the machine and the effects of rotation can be modelled in several ways depending on the complexity of the geometry.

If the rotor is fully axi-symmetric with a smooth surface, then the effects of rotation can be modelled simply by designating the rotor surface as a moving wall. The shear stresses generated in the fluid at the rotor surface will then produce the flow around the rotor. However, it is rare that a rotor can be modelled in this way, except perhaps for some permanent magnet machines.

When the rotor has geometrical features such as a salient pole construction or wafter blades on the rotor, then these features dominate the flows around the rotor and must be properly modelled. If the stator has an end winding that is largely axi-symmetric then the multiple reference frame technique can be used to model the effects of rotation. In this technique, the rotating parts are represented in a reference frame that is given a rotational speed and the stationary parts are modelled in a fixed reference frame. The rotation is modelled as steady-state and the flow generated represents the steady state or mean flow with the rotor at a fixed position relative to the stator. This means that flows resulting from features on the rotor such as wafter blades or salient poles only show an effect on a local region of the stator and not all the way round the stator as would be expected as the rotor moves. So if the average rate of heat transfer on the stator was required, this would have to be estimated by averaging the local values of heat flux around the stator to give a mean value.

In reality, a rotor with non-axisymmetric features generates an unsteady flow within a machine and this can only be fully represented using time-dependent modelling of rotation using the sliding mesh technique. In this technique, the rotor is incremented round and a time-dependent flow pattern generated within the machine. The sliding mesh method is, however, very demanding computationally and typically increases the time needed to solve the CFD problem by an order of magnitude. For this reason, the rotating reference frame method is usually used for CFD in electrical machines. Sliding mesh modelling can often be used as a final check to see if modelling a time-dependent gives any further insights or accuracy.

The only time where a sliding mech model is essential is in a case where non-axisymmetric features in the rotor and stator are both so significant that the shape of the fluid-filled region of the machine changes shape significantly as the rotor moves round. An example might occur in a machine with a salient pole rotor and a significant feature in the stator such as a large single vent at only one point around the machine. The flow through this vent would then be significantly affected by the position of the rotor and a full time-dependent modelling approach should be taken. In most situations, the stator may have non-axisymmetric features such stator coils but as these are identical features that are repeated many times around the stator then circumferentially averaging the results from a multiple reference frame model will be a more practical technique that will give a good approximation to the unsteady effects of rotation.

5.2.3.3   Combined conduction and convection heat transfer

In CFD software, as the energy equation is solved simultaneously with the fluid flow equations, the effects of heat conduction can be modelled within the solid regions of an electrical machine. Modelling heat conduction in solids is a straightforward task computationally as there is no fluid movement to be calculated in these regions. In the solid regions, the important requirement is to ensure that appropriate thermal conductivities are specified with appropriate thermal contact resistances at interfaces between solid components. This may however be somewhat problematic as many solid parts of a machine contain laminated materials, for example laminations in the stator and the rotor and the copper wires or strips in the coils. It may not be practical to model each individual copper conductor or iron lamination with thermal contact resistances in between and so the most common way of representing these materials is to consider them as continuous solids with anisotropic thermal conductivity to represent the increase thermal resistances between the laminations or conductors. This is considered in more detail in the next section.

The majority of the heat within a machine is normally generated in the solid regions principally due to the iron and copper losses. This internal heat generation can easily be represented in the CFD model provided that the location and magnitude of the heat generation is known. In conductors, the rate of heat generation is temperature dependent as the resistivity of copper changes significantly with temperature. It may be appropriate to specify the rate of heat generation in the conductors as a function of temperature. Within the iron laminations, it is important to have a good estimation of the location of the losses to ensure that these are properly represented.

5.2.3.4   Dealing with anisotropic conduction in solids

The construction of an electrical machine with stranded copper conductors and iron laminations means that a detailed representation of these elements is required for an accurate heat transfer analysis. However, this may require a very large number of cells in a CFD mesh to resolve the detail and several ways can be found to simplify this by defining effective thermal conductivities that have different values in different directions.

Where there are iron laminations in a stator or rotor, effective thermal conductivities in the plane of the laminations and perpendicular to the laminations can be determined using the method described in Section 4.1.5. CFD software will normally allow the use of orthotropic thermal conductivities in each coordinate direction and so these can be used provided that the laminations are aligned with the coordinate directions in the CFD model. When laminations are being modelled, the effective thermal conductivities can be given by (4.16) and (4.18).

In the windings, the geometry of the copper conductors can be considerably more complex as described in Section 4.1.1 and there are several ways of determining effective thermal conductivities. Simple formulae such as those for laminations described in Section 4.1.5 may be derived. Alternatively, Section 4.1.1, Figure 4.6 gives some simple relationships for an effective thermal conductivity perpendicular the conductors for windings in stator slots. The FEA method can be more computationally efficient than CFD for determining effective thermal conductivities and the use of this method is explained earlier in this chapter in Section 5.1 for a variety of coil configurations in stator slots.

In iron laminations and windings in slots in stators and rotors, the principal directions of thermal conductivity usually align with the coordinate directions in the CFD analysis and so orthotropic thermal conductivities can easily be used. However, in coils of the end windings, there is an added complexity as the curved shape means that the principal directions of thermal conduction vary around the winding. The most accurate method of representing this situation would be to produce a detailed mesh of the windings in which each copper conductor and the insulation in between was modelled in detail. This is not practical in most cases as the computational effort will be too large. Advanced CFD users may be able to write additional software within the CFD code to allow for local variations in thermal conductivity where the principal directions do not align with the coordinate directions. However, as the main thermal resistance between the end winding and air is at the surface of the winding where there is the outer layer of insulation, a simplified analysis can be done by assuming that the end winding coils are a solid conductor with an isotropic thermal conductivity, typically that parallel to the conductors, surrounded by a layer of insulation of lower thermal conductivity.

5.2.4   Examples of application of CFD

CFD modelling is most commonly used for providing data on convective heat transfer coefficients for use in LPTN models of electrical machines. Alternatively, it may be used to provide data on local winding temperatures, where both the convective heat transfer and the conduction within solids is modelled. When only part of a machine is modelled, then there may be uncertainties over boundary conditions at the interface with other parts of the machine that are not being modelled and this is likely to result in uncertainties in the results and validation with experimental results may be required to give confidence. However, in many cases, CFD is used to investigate the effect of changes in machine design and experimental results may well exist to provide validation of initial CFD models. The changes that result from design changes can then be used with more confidence.

CFD is also used to model the windage loss generated in a machine and this can include the contributions from fan, rotor, and any stirring devices, for example, wafters placed on the end of the rotor to enhance heat transfer in the end region of a machine. There will also be windage losses in the bearings, however, these cannot be modelled in CFD and other sources of data, for instance proprietary information from bearing manufacturers may be used. A good discussion of the use of CFD in modelling windage losses in an electrical machine is provided in Ref. [16].

Two examples are given here where CFD has been used to model part of an electrical machine.

5.2.4.1   Optimisation of the location of a stator vent

An investigation into the optimal location for a single radial stator vent in a medium sized, 4 pole synchronous machine is described in Ref. [18]. The objective was to identify the optimal axial location of a single vent to give the maximum reduction in stator winding temperature. The geometry considered is shown in Figure 5.19. The machine was ventilated from one end and air flows between the rotor and the stator in the air gap and interpolar space as well as along the back of the core in the barrel gap. In this configuration, the air flow distribution in the machine and hence convective heat transfer in the stator is dependent on the vent location and CFD is required to model this complex air flow. A conjugate heat transfer analysis was undertaken including heat conduction within the stator windings and core and heat generation from the iron and copper losses.

The CFD modelling was done using ANSYS Fluent [11] and optimization algorithms within ANSYS Workbench were used to identify the optimum vent location to minimize the stator winding temperature. Figure 5.20 shows how the maximum stator winding temperature varies with the location of the vent. The vent location is expressed in terms of the distance from the inlet end of the stator as a fraction of the core length. The temperature reduction is expressed in terms of a reduction in maximum temperature rise within the stator relative to a base case in which there was no stator vent.

The results showed that the optimum location of the vent was a fraction of 0.7 along the core length from the inlet with a 9.7 K reduction in the maximum winding

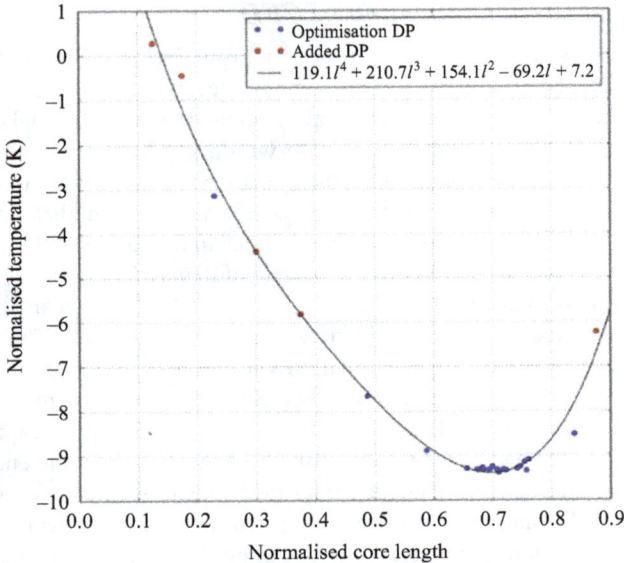

Figure 5.20   The variation in maximum stator winding temperature with axial location of the stator vent. The points represent the results from a CFD calculation and the solid line is a curve fit [18]

temperature. It was also found that the average winding temperature reduced by 8.6 K. It can be seen from Figure 5.20 that the temperature reduction is roughly the same for vent locations from 0.65 to 0.75 and so temperature reduction is not very sensitive to vent location. An interesting feature of the analysis was that if the vent was too close to the inlet, the maximum winding temperature rose above that of the non-ventilated case. This was attributed to air being short-circuited from the air gap to barrel gap close to the inlet end of the stator, reducing the ventilation and cooling in the air gap.

5.2.4.2   Cooling of end windings

An investigation of the arrangement of the end windings in a 10 kW low-voltage TEFC induction motor is described in Ref. [16]. The air flow and convective heat transfer around the end windings of an electrical machine is complex and, in small TEFC machines, the end winding configuration affects not only the convective cooling of the end winding but also the air flow and cooling of the inside of the machine housing in the end region. The end windings of low-voltage machines are usually wound from wire and each coil has a degree of randomness in which it is unlikely that any two coils are exactly the same. As the coils are placed in the machine, they are bound together to form the end winding shape. The end winding may have some degree of openness and the air flow driven by the rotor may pass through to the region behind the end winding and the end region housing. The degree of openness depends on the winding configuration and the location of insulation materials such as interphase separators.

A simple way of representing the shape of the end winding in a CFD model is shown in Figure 5.21(a). However, this is not realistic in representing the uneven surface of the end winding with individual coils and the spaces between them. An alternative way in which each individual coil is represented and blended together to give a more realistic shape, which includes some degree of porosity where there are gaps in between the coils, as shown in Figure 5.21(b).

Each of these geometries was used in a CFD model to calculate the heat transfer within the end region of the machine. It was found that the use of the more realistic end winding configuration not only increases the convective heat transfer from the end windings to the air, but the more porous end winding structure allowed more air to impinge on the outer housing and increase the heat transfer through the housing. The heat flux on the outer housing is shown in Figure 5.22 for each end winding geometry and the rate of heat transfer from the end windings and to the housing for each geometry is given in Table 5.1.

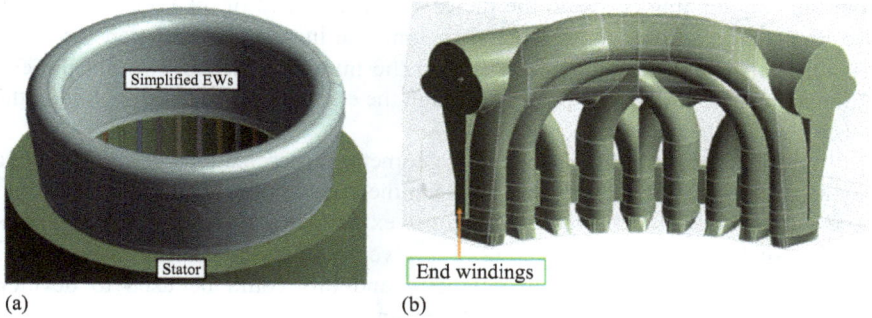

(a)   (b)

Figure 5.21   *(a) Simplified end winding geometry for a low-voltage TEFC machine. (b) More realistic end wind geometry for the same machine in which each individual coil is represented and blended together [16]*

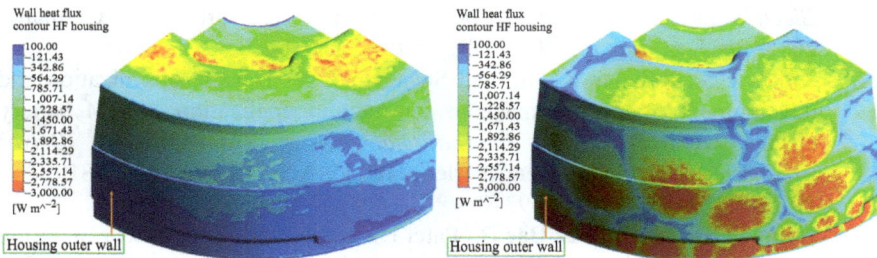

Figure 5.22   *Heat flux contours on the outer wall of the housing with (a) simplified end winding geometry and (b) more realistic end winding geometry [16]*

Table 5.1   Rate of heat transfer from the end winding and to the outer housing with two different end winding configurations [18]

	Rate of heat transfer with simplified end winding geometry (Watts)	Rate of heat transfer with more realistic end winding geometry [Watts]
End winding	17.7	33.7
Outer housing wall	31.1	49.4

Note that the difference between the rate of heat transfer from the end winding and the rate of heat transfer to the outer housing is due to the heat transfer from the machine rotor into the end region.

It can be seen that the more realistic end winding geometry gives an increase of 62% in the rate of heat transfer from the end winding and an increase of 46% in the rate of heat transfer to the outer machine housing. The increase in heat transfer from the end winding is due to the increased surface area of the end winding and also improved air flow within the end region. The increase in heat transfer to the machine housing is attributable entirely to the improved air flow within the end region and particularly to air passing through the end winding and impinging on the housing wall.

In reality, neither of the end winding geometry representations described above may accurately model heat transfer between the end windings and the housing and some validation would be required against experiment. However, in a complex situation such as this, CFD can be used to give a realistic indication of the trends that can result from changes in geometry and this could be used to develop improved end winding configurations to enhance machine cooling.

References

[1]   Bönig J., Bickel B., Spahr M., Fischer C., and Franke J. 'Simulation of orthocyclic windings using the linear winding technique'. *5th International Electric Drives Production Conference (EDPC)*, Nuremberg, Germany; 2015, pp. 1–6. doi: 10.1109/EDPC.2015.7323201

[2]   Jurkovic S., Rahman K. M., and Savagian P. J. 'Design, optimization and development of electric machine for traction application in GM battery electric vehicle'. *IEEE International Electric Machines & Drives Conference (IEMDC)*, Coeur d'Alene, ID, USA; 2015, pp. 1814–1819. doi: 10.1109/IEMDC.2015.7409310

[3]   Jurkovic S., Rahman K., Bae B., Patel N., and Savagian P. 'Next generation chevy volt electric machines; design, optimization and control for performance and rare-earth mitigation'. *IEEE Energy Conversion Congress and Exposition (ECCE)*, Montreal, QC, USA; 2015, pp. 5219–5226. doi: 10.1109/ECCE.2015.7310394

[4]   Momen F., Rahman K. M., Son Y., and Savagian P. 'Electric motor design of general motors' Chevrolet Bolt Electric Vehicle'. *SAE Int. J. Altern. Powertrains.* 2016; 5(2): 286–293.

[5]   Sano S., Yashiro T., Takizawa K., and Mizutani T. 'Development of new motor for compact-class hybrid vehicles'. *World Electr. Veh. J.* 2016; 8(2): 443–449.

[6]   Popescu M. and Dorrell D. G. 'Proximity losses in the windings of high speed brushless permanent magnet AC motors with single tooth windings and parallel paths'. *IEEE Trans. Magn.* 2013; 49(7): 3913–3916.

[7]   Mellor P., Wrobel R., and Simpson N. 'AC losses in high frequency electrical machine windings formed from large section conductors'. *IEEE Energy Conversion Congress and Exposition (ECCE)*, 2nd edn.; 2014, pp. 5563–5570.

[8]   Sullivan C. R. 'Optimal choice for number of strands in a Litz-wire transformer winding'. *IEEE Trans. Power Electron.* 1999; 14(2): 283–291.

[9]   Litz Wire Types and Constructions – A description of the various styles of Litz wire – Type 1 through Type 8 [Online]. Available from https://www.newenglandwire.com/product/litz-wire-types-and-constructions/ [Accessed 27 July 2020].

[10] Versteeg H. K. and Malalasekera W. *An Introduction to Computational Fluid Dynamics – The finite volume method*, 2nd edn.; 2006.

[11] Ansys Fluent Fluid Simulation Software [Online]. Available from https://www.ansys.com/products/fluids/ansys-fluent [Accessed 12 July 2020].

[12] Computational Fluid Dynamics (CFD) Software Program Solutions [Online]. Available from https://www.ansys.com/products/fluids/ansys-cfx [Accessed 12 July 2020].

[13] Pickering S., Wheeler P., Thovex F., and Bradley K. 'Thermal design of an integrated motor drive'. *IECON 2006 – 32nd Annual Conference on IEEE Industrial Electronics*, Paris, France, 2006, pp. 4794–4799, doi: 10.1109/IECON.2006.348109.

[14] Connor P. H., Pickering S. J., Gerada C., *et al.* 'Computational fluid dynamics modelling of an entire synchronous generator for improved thermal management'. *IET Electr. Power Appl.* 2013; 7(3): 231–236, 3, doi: 10.1049/iet-epa.2012.0278.

[15] Connor P. Ph.D. Dissertation, University of Nottingham, Nottingham; 2014.

[16] Rocca S. L., Pickering S. J., Eastwick C. N., Gerada C., and Rönnberg K. 'Fluid flow and heat transfer analysis of TEFC machine end regions using more realistic end-winding geometry'. *IET, J. Eng.* 2019; 2019(17): 3831–3835.

[17] Pickering S. J., Lampard D., Mugglestone J., Shanel M., and Birse D., 'Using CFD in the design of electric motors and generators'. Paper published in 'Computational Fluid Dynamics in Practice' Ed by Norman Rhodes, Published by Professional Engineering Publishing, 2001. ISBN 1 86058 352 0.

[18] Bersch K., Connor P. H., Eastwick C. N., Galea M., and Rolston R. 'CFD optimisation of the thermal design for a vented electrical machine'. *IEEE*

Workshop on Electrical Machines Design, Control and Diagnosis (WEMDCD), Nottingham, UK; 2017, pp. 39 D. 44, doi: 10.1109/WEMDCD.2017.7947721.

[19]   Gerling D. 'Concentrated windings'. In: *Electrical Machines. Mathematical Engineering*, vol. 4. Berlin, Heidelberg: Springer; 2015, pp. 449–462.
[20]   Bersch K. Ph.D. dissertation, University of Nottingham, Nottingham; 2019.
[21]   Micallef C., Pickering S. J., Simmons K. A., and Bradley K. J. 'Improved cooling in the end region of a strip-wound totally enclosed fan-cooled induction electric machine'. *IEEE Trans. Ind. Electron.* 2008; 55(10): 3517–3524, doi: 10.1109/TIE.2008.2003101.

Chapter 6

Thermal test methods

In this chapter, Section 6.1 looks at the most common devices and methods used to measure temperature, a heat flux, and air flow. The advantages of the different methods are highlighted. Section 6.2 shows a test method that can be used to measure winding anisotropic thermal conductivity. As an example, typical values of the thermal conductivity through the wire-to-wire insulation system to the slot wall and along the conductors to the end-winding are given for different sizes of copper and aluminum conductors with varnish and epoxy resin impregnation material. Also for compressed windings. Section 6.3 gives some basic details of test methods that can be used to measure the different components of loss in the machine. Segregation of losses is a large topic in itself so this is only covered in basic detail here and reference is made to different international standards commonly used for the measurement of losses in electrical machines. The measurement of windage losses in more detail is described as this relates to air flow. In Section 6.4, guidance is given on how best to calibrate a thermal model using test data. This can be very useful to increase model accuracy and provide a useful insight of how a machine compares with other similar machines in terms of manufacturing goodness and quality of the design. Finally, in Section 6.5, details are given of a relatively new test method in which a short thermal transient test is used to estimate the winding-to-stator thermal resistance and winding thermal capacitance.

6.1 Measurement methods

6.1.1 Temperature measurement

The most common methods for temperature measurement in electrical machines are thermocouples and resistance temperature detectors. Surface temperatures can also be measured by infra-red thermography and liquid crystal paints.

6.1.1.1 Thermocouples

These are thermoelectric devices that operate using the Seebeck effect whereby a circuit made from two dissimilar conducting metals will generate an EMF if the two junctions between the metals are at different temperatures, as illustrated in Figure 6.1. The voltage generated by the temperature difference depends on the materials used in the conductors and there are standard pairs of metals that are used

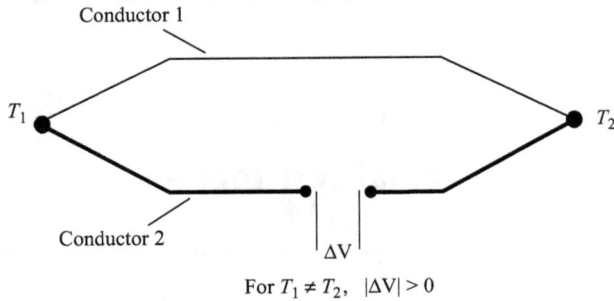

Figure 6.1   *Circuit illustrating the thermocouple principle*

Table 6.1   *Characteristics of common thermocouples*

Thermocouple type	Material pair	Useful temperature range	Typical output (μV/°C)
Type K	Nickel–chromium/nickel-Aluminum	−200 to +1260	~ 41
Type T	Copper/copper–nickel	−200 to +200	~ 48
Type J	Iron/copper–nickel	−40 to +750	~ 52

for thermocouples. Further information is given in Ref. [1]. These standard metal pairs are well characterized in terms of the EMF generated by the temperature difference and the most commonly used are listed in Table 6.1. Thermocouple specifications are covered by international standards, in particular IEC 60584. Type K thermocouples are the most popular as they work over a wide range of temperatures. Thermocouples measure the difference in temperature between the two junctions in the circuit and this is normally achieved in practice by having one junction located where the temperature is to be measured, usually known as the "hot junction" and the other junction located in the measuring instrument where there is also a "cold junction" reference temperature measurement. The temperature of the hot junction can be determined by reference to the cold junction reference temperature. While thermocouples are very widely used, they have the disadvantage that the connecting wires form the hot junction to the measuring instrument need to be made from the same two conducting materials used at the thermocouple tip, or from materials that have similar characteristics. There is also the need to have a temperature measurement at the cold junction.

Thermocouples are supplied in a variety of physical forms ranging from insulated wires to fully encapsulated systems in metallic sheaths. Thermocouple systems with metallic sheaths are normally the more robust and these may vary in diameter from sub millimetre to over 10 mm. Thermocouples are also available in a variety of specialized arrangements for example for the measurement of rapidly changing air temperatures or surface temperature.

In electrical machines, small diameter thermocouples are often embedded in windings and other parts of the machine, sometimes inserted in purpose made holes. Surface temperature can be measured with proprietary configurations such as those with a circular disc shown in Figure 6.2(a). To obtain the most accurate reading, it is advisable to apply a thermal paste between the probe and the surface being measured to ensure a good thermal contact. Air temperature can be measured using a probe with a small diameter, such as the example also shown in Figure 6.2(b).

In using thermocouples, two important factors must be taken into account. First, it is only the temperature at the junction of the two conductors that is measured. So in the case of thermocouples embedded in a metallic sheath, it is the temperature at the tip that is measured. If temperature is being measured where there is a temperature gradient, for instance close to an interface between two materials, then it is important to ensure that there is good thermal contact with the tip of the thermocouple. This can often be enhanced by using a high thermal conductivity compound between the thermocouple tip and the object being measured. Also if possible the thermocouple should be positioned so that the region near to the tip is in the direction of minimal temperature gradient to reduce errors due to conduction along the leads or sheath. Second, in regions of alternating magnetic field, spurious EMFs may be generated in the thermocouples conductors. These errors can be minimized using a twisted pair of conductors, but small errors can still occur. The magnitude of these errors can be established by momentarily de-energizing a machine to see if temperatures change. The best way of allowing for these errors is to de-energize a machine at a time when a temperature measurement is required and then plotting the cooling curve from the thermocouple measurements. This curve can be extrapolated back to the time the machine was de-energized to determine the temperature at that time.

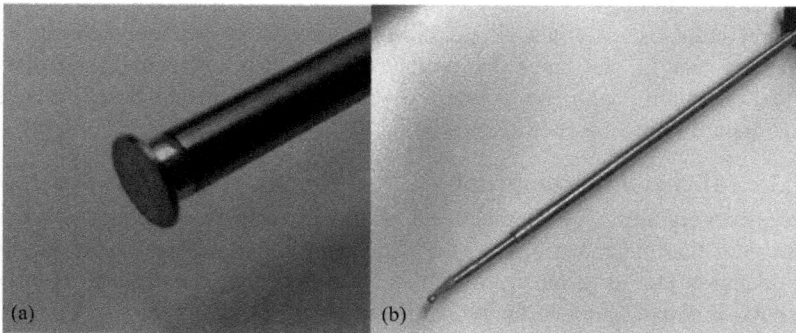

Figure 6.2   Examples of thermocouple probes for surface temperature measurement (a) and air temperature measurement (b)

6.1.1.2   Resistance temperature detectors

These temperature measurement devices work on the principle that the resistance of a conductor is temperature dependent. For metals, resistance increases with temperature in accordance with the following equation:

$$R(T) = R_0[1 + a_1(T - T_0) + a_2(T - T_0)^2] \tag{6.1}$$

where $R(T)$ is the resistance at temperature T, R_0 is the resistance at temperature T_0, and a_1 and a_2 are coefficients determined experimentally for particular metals.

For a small range of temperature, the higher order term may be omitted and the equation becomes:

$$R(T) = R_0[1 + a_1(T - T_0)] \tag{6.2}$$

In this linear equation, a_1 is generally known as the temperature coefficient of resistance.

This phenomenon is used in electrical machines to determine the average temperature in a winding by measuring the change in resistance of the copper conductors. Proprietary resistance temperature detectors (RTD) are available using platinum as the resistance material and their use is described by the international standard IEC 60751. The most common platinum resistance thermometers are known as PT 100 and PT 1000. The PT 100 has a resistance of 100 ohms at 0 °C, while the PT 1000 has a resistance of 1,000 ohms at the same temperature. These resistance thermometers have a very good linearity with the temperature. Platinum resistance thermometers have temperature coefficient of resistance (a) values from 0.00375 to 0.003928 [/°C].

The advantage of RTDs is that the connecting leads to the instrument do not need to be made from a special material and so copper can be used. However, the resistance of the leads can affect the accuracy of the measurement and the most accurate devices use a 4 wire system where one pair of wires is used to compensate for the resistance in the sensor leads. With RTDs, there is also no need for a reference temperature to be measured in the measuring instrument. A disadvantage of RTDs is that they are larger than thermocouples and so are more difficult to embed within windings.

Semiconductor materials also exhibit temperature-dependent resistance and they have a specific resistance that is much higher than metals and the resistance falls exponentially with temperature. Temperature sensors based on semiconductors are known as thermistors.

6.1.1.3   Infra-red thermography

Surface temperatures can be measured using infra-red thermography. All substances emit thermal radiation and the maximum radiation heat flux from a perfectly black surface is given by (2.68) in Section 2.3. All real surfaces have an emissivity less than 1 and so the radiant heat flux q'' is given by:

$$q'' = \varepsilon \sigma T^4 \tag{6.3}$$

where ε is the emissivity of the surface, σ is the Stefan–Boltzmann constant (see Section 2.3), and T is the absolute temperature of the surface. Infra-red cameras are sensitive to long wavelengths of thermal radiation and by measuring the radiant heat flux from a surface can determine the temperature. Temperature measurement can either be done using single-point measurement devices, that give the mean temperature over a small area, or multi-pixel cameras that give a thermal image. Thermography is useful for surface temperature measurement but a direct line of sight is needed between the camera and the surface. Some small infra-red temperature detectors, sometimes known as infra-red thermocouples, are available which can be positioned within equipment to measure surface temperature in inaccessible locations.

The main uncertainty with infra-red thermography is the emissivity of the surface. Most oxidized metal surfaces, non-metallic or painted surfaces have high emissivities typically about 0.95. However, polished metal surfaces and in particular, aluminum, even when oxidized, have emissivities that may be less than 0.2. The radiant heat flux from these low emissivity surfaces is therefore much less than a high emissivity surface and must be allowed for. As aluminum is often used in electrical machines, errors may occur if thermography is used. There are two ways of avoiding these errors. First, the emissivity of the material surface can be calibrated with an alternative method of surface temperature measurement. Second, the surface could be painted with a high emissivity paint and this is preferable if possible. One of the problems with infra-red temperature measurement of low emissivity surfaces is that not only do these surfaces emit less radiation but they also reflect radiation that is incident upon them. So, radiation from a nearby hot surface could be reflected from a low emissivity surface again giving a temperature measurement error. Since the 2020 Covid-19 pandemic, infra-red temperature surface temperature measurement devices are now widely available as they are regularly used in health care applications for measuring body temperature.

Figure 6.3 shows an example of a motor under test that has been photographed with a thermal camera. The motor is a 10 kW, 40 V six-phase induction motor for an electric van application. It has water jacket cooling.

6.1.1.4   Liquid crystals

Temperature sensitive liquid crystals are available in the form of proprietary thermochromic paints or inks that change colour with temperature. These are available to change colour over different temperature ranges and may be appropriate for surface temperature measurement in some applications.

6.1.2   Heat flux measurement

Sensors are available for measuring the heat flux from a surface and these may be used, combined with a surface temperature measurement, to calculate the convective heat transfer coefficient (h) on a surface using the equation:

$$h = \frac{q''}{(T_{surface} - T_{fluid})} \tag{6.4}$$

where q'' is the heat flux. These proprietary heat flux sensors work by measuring the temperature difference across a surface through which the heat flux is passing.

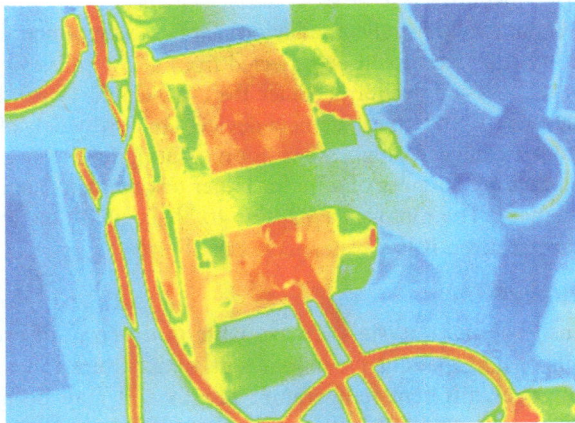

Figure 6.3   Thermal camera image of an electric van 10 kW water cooled 6-phase induction motor under test

They are usually made from a thin plastic film that may be < 1 mm in thickness with an array of fine wire thermocouples located on both sides of the film to provide a voltage that is proportional to the temperature difference. The heat flux can be calculated from Fourier's Law:

$$q'' = k\left(\frac{\Delta T}{\Delta x}\right) \tag{6.5}$$

where Q'' is the heat flux, k is the thermal conductivity of the plastic film of thickness Δx, and ΔT is the temperature difference across the film.

Heat flux sensors are manufactured to be small and thin so that they do not disrupt the boundary layer on the surface which would give a false indication of heat transfer coefficient and also so that they do not add a significant thermal resistance to the surface. They usually have a thermocouple included on the upper surface to measure surface temperature at the same time. When using heat flux sensors, it is important that they are in good thermal contact with the surface so that they measure the true surface temperature. It is also helpful to install them in such a way that the leads are downstream of the sensor so as not to disrupt the boundary layer and give an artificially high heat transfer coefficient.

6.1.3　Air flow measurement

6.1.3.1　Bulk air through flow measurement through a machine

Bulk air flow through an open ventilated machine or through the fan of a TEFC machine can be measured by positioning the motor so that the air flow can be drawn through a duct air flow measuring device. Such an arrangement is shown in Figure 6.4 where the machine is positioned to take in from a plenum and discharge to atmosphere. A calibrated inlet duct with a conical inlet manufactured to British Standard BS848 is used to measure the air flow entering the plenum [2]. Care must be taken to size the air flow measurement duct so that it imposes a minimum back pressure on the motor inlet, to ensure that the air flow through the fan is not reduced.

Air flow measurement ducts manufactured to BS 848 are easy to make and convenient to use for air flow measurement in electrical machines. They impose a small pressure loss and the air flow is calculated based on the pressure difference

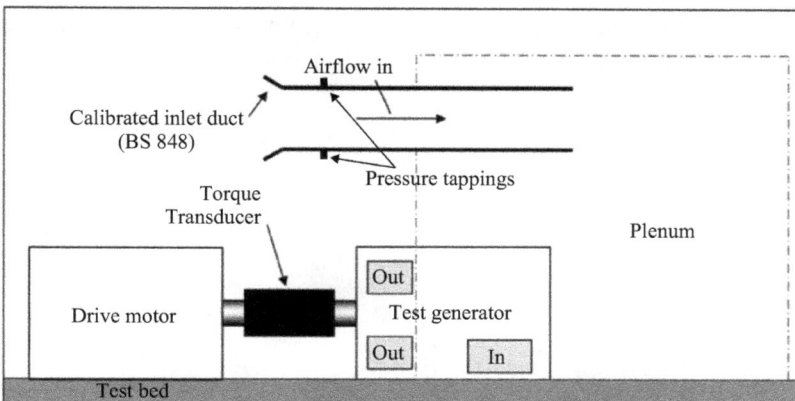

Figure 6.4　Experimental test arrangement for measuring bulk air flow through a machine (adapted from Ref. [2])

between the atmosphere and the pressure tappings located half a diameter down-stream from the conical inlet. If a machine has only one inlet (e.g. TEFC motor), it may be possible to position the duct directly against the machine, but it is advisable to insert some honeycomb flow straightener towards to outlet end of the duct to prevent air flow swirl due to the fan, at the duct inlet, which would affect the calibration.

6.1.3.2   Air velocity measurement – pitot probes

Pitot probes measure air velocity by converting the velocity or dynamic pressure in moving air to give a total pressure at the stagnation point immediately upstream of the probe inlet. This can be compared with the local static pressure to determine the air velocity as shown in the equation:

$$p_t = p_s + \frac{1}{2}\rho V^2 \tag{6.6}$$

giving

$$V = \sqrt{\frac{2(p_t - p_s)}{\rho}} \tag{6.7}$$

where p_t and p_s are the total and static pressures in the air flow at the measurement point, respectively, V is the air velocity, and ρ is the air density. The equations above will give accurate values of velocity for low-speed air flows to within 0.25%, where the velocity is less than 10% of the speed of sound and the pressures can be accurately measured. For air velocities or more than 30% of the local speed of sound, the error will be in excess of 2% and other correction factors will be needed, as explained in Ref. [3]. Note that the speed of sound in air varies as the temperature varies from 330 to 390 m/s over the temperature range from 0 °C to 100 °C.

A simple probe will just measure the total pressure and requires an alternative measurement of the static pressure in the air flow. Pitot-static probes combine the total and static pressure measurements in one probe. Both of these probes are shown in Figure 6.5.

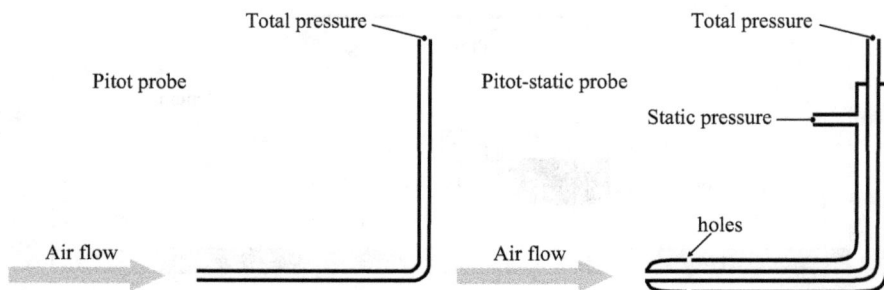

Figure 6.5   Pitot probes for air velocity measurement

The pitot probes shown in Figure 6.5 must be pointing directly into the flow and errors will occur if the flow approaches at an angle or if the probe is positioned too close to a wall. Similarly, if pitot probes are used to measure velocities in ducts, the cross-sectional area of the probe must be much less than the cross-sectional area of the duct to reduce the blockage effect as the air velocity increases to get past the probe. For example, if the cross-sectional area of the probe is less than 1% of the duct, then the velocity error will be less than 2%. More detail of the use of pitot probes is given in Ref. [3].

Pitot probes are very useful for measuring air flow when air flow direction is known and the probe can be positioned the face the flow. However, in some cases, in electrical machines, the air flow may be swirling and the direction unknown. Specialized multi-hole pitot probes can be used in these circumstances that can be calibrated and used to give air velocity measurements when the air flow direction is not known [3].

6.1.3.3   Hot wire air velocity probes

Hot wire anemometry is a technique for velocity measurement widely used in experimental fluid mechanics for taking detailed flow measurements in gases [3]. The operating principle is to expose a wire, which is heated by an electric current, to an air flow. If the wire is made from a material with a temperature coefficient of resistance, then the resistance will change as the temperature changes due to the convection from the air flow. The hot wire probe can then be calibrated to deter-mine the air velocity from the change in resistance or current in the wire. There are three methods of using these probes depending on how the probe is controlled: constant current, constant voltage, or constant temperature probes. When probes using very fine wire are used, very detailed measurements can be made to inves-tigate the variation in velocity with location and also with time as the probe will have a very small thermal mass and the fluctuations in velocity in turbulent flows can be measured. These probes are usually only appropriate for measurements in carefully controlled laboratory conditions. Examples of fine wire probes are given in Figure 6.6. More rugged probes are also available for more general velocity

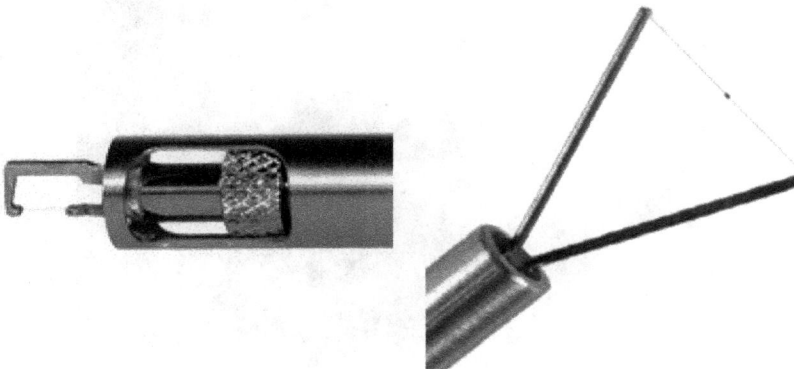

Figure 6.6   Examples of fine wire hot wire anemometer probes

measurement where the hot wire element can be ~1 mm in diameter and ~10 mm long. These are calibrated for general air velocity measurement and can easily be used in electrical machines, provided that there is access for the probe to the location where air velocity is to me measured.

6.1.3.4   Rotating vane air velocity probes

In this type of anemometer, the air speed is measured by the speed of the rotating vane. They are better for higher velocity flows and lower air speeds are better measured using hot wire anemometers. The use of the rotating vane anemometer is easy but the main drawback is their relatively large size. Figure 6.7 shows an example of a rotating vane anemometer. Depending on the size of the probe, the measuring range is typically from 0.4 to 30 m/s, with a resolution of about 0.1 m/s.

6.1.3.5   Laser-based methods

There are several techniques available for measuring air velocity without the need to insert a probe inside a machine. These are laser based techniques, such as Laser Doppler Anemometry (LDA) and Particle Imaging Velocimetry (PIV), in which the movement of particles within a flow is measured [3]. These techniques require optical access for lasers within the machine and this may require the manufacture of special windows for the laser beam(s). While these techniques are able to provide very accurate, high-resolution, and high-frequency response measurements, they require expensive equipment and normally are only done under carefully controlled laboratory conditions.

6.1.4   Liquid flow measurement

In electrical machines that are cooled by water or other liquids, the liquid flow rate will normally be measured in the liquid supply pipes as a volume flow rate rather

Figure 6.7   Example of a rotating vane anemometer

than a velocity. Venturi meters or orifice plate meters are commonly used for liquid and gas flow measurement in pipes and ducts and there are standards that define their configuration and calibration (e.g. ISO 5167 and BS 1042). In these devices, the flow rate is related to the variation in static pressure within the flow as it passes through a venturi tube or an orifice placed in a pipe. Many other proprietary flow meters are also available in which the flow may be detected by a rotating vane or turbine. Ultrasonic techniques may also be used. The choice of the most suitable meter has to be based on the application and the required accuracy.

6.2   Winding insulation system thermal conductivity

The accurate estimation of winding thermal behaviour is critical to avoid the failure of the winding insulation system. For computationally efficient thermal analysis, the winding is commonly modelled as a composite material with anisotropic thermal conductivity, as discussed in Section 4.1.1. The equivalent thermal conductivities that are applicable to a specific direction are dependent on the conductor geometry, thermal properties of constituent materials and even manufacturing factors. An extensive range of experimental measurements have been carried out to calibrate the equivalent thermal conductivities of the windings predominantly based on conventional round conductors [4]. The empirical studies have provided very useful insight into the influence of wire fill factor and thermal properties of the insulation material on the winding equivalent thermal conductivity. Moreover, the winding is usually impregnated with epoxy resin and different impregnation process (e.g. trickling, dipping, dip-rolling, vacuum pressure impregnation, etc.) will lead to different impregnation goodness. Experimental methods are a useful way to examine how the impregnation process affects the impregnation goodness and thus the equivalent thermal conductivity. Therefore, due to changes in manufacturing process, insufficient material data, and new conductor shapes, experimental measurements are required to calibrate the winding model for accurate prediction of winding temperature rise.

Basically, the experimental setup to emulate the equivalent thermal conductivity of a winding sample is straightforward based upon the principle of heat conduction (Fourier's law). The one-dimensional thermal conductivity is obtained by applying a known heat flux (q'') to the winding sample and measuring the temperature difference (T) along the direction of heat transfer for a specific length (l):

$$k = \frac{q''l}{\Delta T} \tag{6.8}$$

To generate the heat flow path in the direction of interest as shown in Figure 6.8, the winding sample is heated by a resistive heater energized from a DC source. To ensure a uniform heat flow across the winding sample from the heater, a thermally conductive hot plate is placed between the heater and winding sample. At the other end, the winding sample is cooled by a water-cooled cold plate in order to

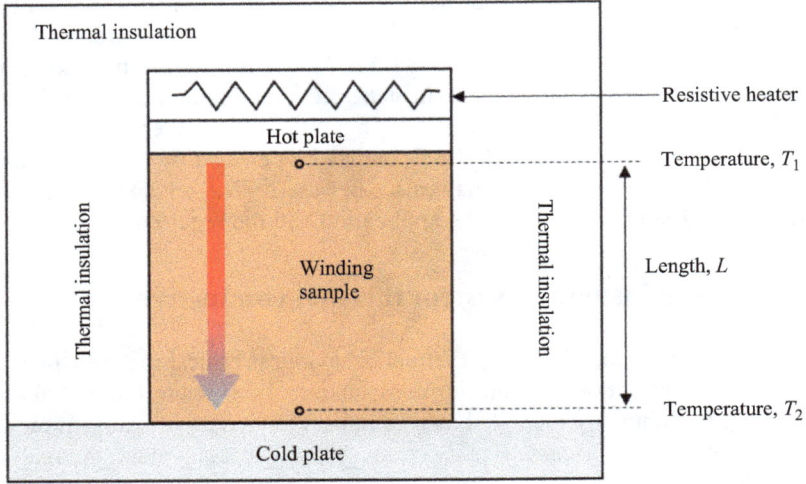

Figure 6.8   Experimental setup for the thermal conductivity measurements

create a temperature gradient to allow accurate measurement of thermal conductivity. The remaining outer surfaces of the winding sample are thermally insulated.

Based on the experimental setup, three winding samples constructed with round conductors with different material configurations were examined [4]. One winding sample is comprised of copper conductors with polyamide-imide insulation of thermal conductivity 0.26 W/m/K. The other one is similar but with aluminum conductors instead. The last sample is comprised of aluminum conductors with aluminum oxide (Al_2O_3) insulation of thermal conductivity of 2 W/m/K. The samples are vacuum impregnated with epoxy resin of conductivity 0.85 W/m/K. In the experimental study, the equivalent thermal conductivities of those winding samples are strongly affected by the slot fill factor due to different winding types as described in Section 5.1. For uniformly distributed round conductors, the thermal conductivity across the conductors can usually assumed to be isotropic. However, experimental measurements in different axes are suggested for non-uniform conductor distribution.

The use of Litz wire to reduce the additional losses in the winding due to skin and proximity effects for high-frequency machines is getting more and more popular [5]. However, the construction of Litz wire is different from conventional round wire and is basically constructed by having a bunch of individually insulated strands of finer wires twisted together. Some Litz wire is further compressed into a rectangular profile for a better slot fill factor. Some Litz wire might even be formed by several insulated Litz wires twisted together. Because of the complexity of Litz wire construction, the analytical method as described in Section 4.1.1 used to approximate the equivalent thermal conductivity of the winding is no longer suitable for accurate predictions. In Ref. [6], a winding sample constructed with Type 8 Litz wire was tested by using a similar experimental setup to Figure 6.8. The measured thermal conductivities vary depending on which axis is measured. The thermal conductivity ranges from 1.2 to 1.6 W/m/K across the Litz wire, while the thermal conductivity along the winding is much higher at 167.3 W/m/K approximately.

For windings with profiled rectangular conductors, anisotropic thermal conductivities need to be considered as the winding homogenization technique applied to conventional round conductors, assuming a single equivalent thermal conductivity across the conductors regardless of heat transfer direction, is no longer valid. The experimental measurements of thermal conductivity of winding samples with profiled rectangular conductors has been presented in Ref. [7] by using conductors with a variety of aspects ratio as given in Figure 6.9 and Table 6.2. To be representative, the winding samples are prepared in accordance with the manufacturing techniques and materials commonly founded in electrical machine construction. All the winding samples are vacuum impregnated with either solvent based varnish or thermally conductive epoxy resin. All the conductors used in the sample have the same grade of polyamide-imide enamel coating (class N) with thermal conductivity of 0.26 W/m/K. The measurement data shows that the difference between thermal conductivity in the x-axis (k_x) and the y-axis (k_y) is larger for conductors with increased aspect ratio, i.e. winding sample e with an aspect ratio of 8.3 when compared to winding sample a with an aspect ratio of 5.0. By comparing winding samples c and d, low conductor fill factor has a negative impact on the thermal conductivities. With similar conductor fill factor, the winding sample impregnated with epoxy resin exhibits much higher thermal conductivity values as the epoxy resin has higher thermal conductivity than varnish which is

Figure 6.9   Winding samples with profiled rectangular conductors of different aspect ratio. (a)–(e) are copper conductors except (b) is aluminum conductors [7].

Table 6.2 Measured thermal conductivities of winding samples with profiled rectangular conductors of different aspect ratios [7]

Winding sample	Conductor material	Conductor fill factor (%)	Impregnating material	Conductor dimensions [mm x mm]	Conductor aspect ratio	k_x (W/m/K)	k_y (W/m/K)
a	Copper	75	Varnish	1.4 × 7	5.0	2.0	2.2
b	Aluminum	77	Varnish	1.4 × 7	5.0	1.9	2.0
c	Copper	27	Epoxy resin	1.5 × 3	2.0	2.0	2.3
d	Copper	73	Epoxy resin	1.5 × 3	2.0	4.0	4.3
e	Copper	77	Varnish	1.2 ×10	8.3	1.9	3.1

Figure 6.10 Tooth wound coil with compressed aluminum-stranded winding

0.65 W/m/K. On the other hand, a copper conductor shows slightly higher thermal conductivities than an aluminum conductor.

For a bobbin wound coil of the shape shown in Figure 6.10, the thermal conductivity to the slot wall has been measured to be 0.76 W/m/K [8]. This was for an impregnated aluminum winding with round conductors. In order to increase the thermal conductivity, the coil can be pressed under high pressure to form a compressed winding as shown in Figure 6.10. The compressed winding in this case has a thermal conductivity of 2.32 W/m/K. The compressed winding has a very high slot fill with the round conductors deformed to minimize the coil cross-sectional area. It is surprising that with the deformation the wire insulation is not compromised. On top of the increase in thermal conductivity, the compressed winding also benefits from a smaller overall coil size.

The thermal modelling of a full electric machine normally utilizes composite materials with anisotropic thermal conductivity to represent the winding region rather than a detailed physical representation. Therefore, the winding equivalent thermal conductivities calibrated through experimental measurements can provide confidence to the machine designer in giving accurate estimation of the thermal behaviour of the winding.

6.3 Motor losses

In order to predict the temperature distribution in an electrical machine, it is important to accurately know both the magnitude and the distribution of the power losses in the machine. Testing of the losses in the machine can help with this but does have some pitfalls. It is impossible to test each loss mechanism in isolation and methods of loss separation are required. For example, a no-load test procedure is used to estimate the mechanical losses and core losses in induction motors. The mechanical losses are estimated by extrapolating the total loss curve to a value where the current/voltage is zero. The core losses will be determined by subtracting the previously determined mechanical losses and the calculated DC copper losses (using measured current and resistance values) from total losses.

Similarly, an open-circuit test procedure is used to estimate the core losses in a brushless permanent magnet (PM) along with another open-circuit test procedure using a non-magnetized rotor to estimate the mechanical losses.

Certain loss components are practically impossible to measure, i.e. magnet losses, retainers, and sleeves losses or AC winding losses. Only through calculations can the values of such complex losses be estimated.

Other complications include how much of the total iron loss is in the tooth and back iron (electromagnetic calculations can help here); how much loss is on the stator and how much on the rotor; predicting the magnitude of AC losses in the winding and at which particular points these are concentrated in the slot and end-windings (electromagnetic calculations can help here).

International Standards for measurement of losses and efficiency are useful when setting up methods for loss segregation procedures. Examples of EU (IEC) and USA (IEEE) standards for various machines are briefly reviewed below (note that other international standards such as JEC from Japan are also available):

IEC 60034-2-1:2014: Rotating electrical machines – Part 2-1: Standard methods for determining losses and efficiency from tests (excluding machines for traction vehicles). This standard applies to DC machines and AC synchronous and induction machines.

IEC 60034-2-2:2010: Rotating electrical machines – Part 2-2: Specific methods for determining separate losses of large machines from tests. This supplement to IEC 60034-2-1 applies to large rotating electrical machines in which full-load testing is not practical. It establishes additional methods of determining separate losses and to define an efficiency.

IEC 60034-2-3:2013: Rotating electrical machines – Part 2-3: Specific test methods for determining losses and efficiency of converter-fed AC induction motors. Specifies test methods for determining losses and efficiencies of converter-fed AC induction motors that form part of a variable frequency power drive system. Also gives details on the additional harmonic losses. This standard has recently been replaced by IEC 60034-2-3:2020.

IEC 60034-2-3:2020: Rotating electrical machines – Part 2-3: Specific test methods for determining losses and efficiency of converter-fed AC motors. Specifies test methods and an interpolation procedure for determining losses and efficiencies of converter-fed motors. The document also specifies procedures to determine motor losses at any load point (torque, speed) within the base speed range (constant torque range, constant flux range) based on the determination of losses at seven standardized load points.

IEC 60349-4:2012: Electric traction – Rotating electrical machines for rail and road vehicles – Part 4: PM synchronous electrical machines connected to an electronic converter. This standard applies to converter-fed PM synchronous motors or generators (machines) forming part of the equipment of electrically propelled rail and road vehicles. The object of this part is to enable the performance of a machine to be confirmed by tests and to provide a basis for assessment of its suitability for a specified duty and for comparison with other machines. Particular attention is drawn to the need for collaboration between the designers of the machine and its associated converter as detailed in this standard.

IEEE Std 112-2004: IEEE Standard Test Procedure for Polyphase Induction Motors and Generators. This standard provides the basic test procedure for evaluating the performance of a polyphase induction motor or generator of any size. Each revision of the standard since its 1964 introduction as an IEEE standard has been to keep the standard current with improvements in instrumentation, with improvements in test techniques, with increased knowledge in the art of measurements, and with the constant change in the needs and desires of the machine users and of those concerned with energy conservation and the like. Major portions of the document have been rearranged to accomplish this and the user is cautioned to check any external references to particular clauses of previous versions for the correct clause number in this version. Each individual test is defined and each efficiency test method is now covered in more detail and step-by-step instructions are presented.

IEEE Std 112-2017: IEEE Standard Test Procedure for polyphase induction motors and generators.

IEEE STD 114-2010: IEEE Standard Test Procedure for Single-Phase Induction Motors

IEEE 1812-2014: Trial-Use Guide for Testing PM Machines. This trial-use guide contains instructions for conducting tests to determine the performance characteristics and machine parameters of PM machines. It is not intended that this guide shall cover all possible tests, or tests of a research nature, but only those general methods that may be used to obtain performance data and machine parameters.

6.3.1 Measurement of windage losses

Windage losses in an electrical machine can be determined either by experimental measurement or by calculation using computational fluid dynamics (CFD). The use of CFD is described in Section 5.2 and has the advantage that windage losses from individual elements of a machine, for example the fan or individual elements of the rotor, and the effect of design changes, can easily be investigated by modelling. There are two methods in which the windage losses of a machine can be determined experimentally as follows:

- *Torque transducer* – A range of proprietary torque transducers are available and these usually measure the strain in a shaft by a variety of means. Figure 6.4 shows a torque transducer located between two electrical machines. The windage loss in the driven machine can be measured by shaft torque when the machine is unloaded. Torque transducers will usually measure accurately only over a limited range and as the windage loss in an electrical machine is normally significantly less than 5% of the machine driving torque, it is usually necessary to buy a torque transducer specifically rated to measure these low torques.

- *Torque reaction from shaft-mounted machine* – An alternative means of measuring torque is to mount the machine in bearings on each end of the shaft allowing the motor to swing freely. The reaction torque when the machine is driven can then be measured by torque arm reacting onto a fixed load cell. The load cell would be selected to measure accurately over the torque range encountered. This method for torque measurement usually requires the machine to have a shaft at both ends on which support bearings can be mounted.

The torque measurements made by the methods described above include the contributions from the machine bearings as well as windage within the machine from the rotor and fan. The fan windage loss can be measured by removing the fan to observe the reduction in loss. Most electrical machines use rolling element bearings as these generally provide the lowest friction. For these types of bearing the loss is often a variable quantity dependent on the quantity and temperature of any lubricating grease in the bearing and the presence of any rubbing seals in the bearing. It can therefore be difficult to accurately measure the rotor windage loss. However, it is possible to minimize the bearing loss by removing any grease and rubbing seals from the machine bearings and using a small quantity of low viscosity lubricant instead. The machine may then be run under no load conditions for short durations with minimal bearing loss and this can give a better indication of the rotor windage loss.

6.3.2   Calorimetry for measurement of total loss

As explained above, it is not easy to accurately measure all the losses in an electrical machine and so to determine the machine efficiency the output power can be subtracted from the total input power (e.g. IEEE Standard 112). However, for machines of high efficiency, this requires the subtraction of two large numbers and a high accuracy in each measurement is therefore required. An alternative is to measure the loss directly and this can be done using calorimetry. The concept is to place the electrical machine under test in an insulated enclosure and measure the enthalpy gain in an airflow passing through the enclosure. Once steady state has been reached, the total loss appears as the enthalpy rise in the ventilation flow passing through the enclosure. Calorimeters require very careful design and manufacture to ensure that there is no heat loss by conduction through the enclosure and this may be done by using an active temperature control and heaters to ensure that there is zero temperature gradient across the enclosure structure, including machine supports. It is also essential to measure the air flow and temperature passing through the calorimeter to a high accuracy. A high performance calorimeter is described in Ref. [9]. As such, calorimeters are not tools that can be used widely for machine loss measurement, but they are specialist facilities that can be used to accurately calibrate other loss prediction and measurement tools.

6.4 Thermal model calibration

As discussed in Chapter 4, it has been seen that there are many complex issues (including some manufacturing issues) when setting up an accurate thermal model for an electric machine. Calibration of models using test data helps to increase model accuracy. Also, the model calibration provides useful insight into how the designed machine compares with other machines in terms of manufacturing goodness and quality of design. There are many manufacturing uncertainties that can affect machine temperature rise such as:

- Goodness of effective interface between stator and housing.
- Goodness of effective interface between winding and stator laminations.
- How well the winding is impregnated or potted.
- Leakage of air from open fin channels for TEFC machines.
- Cooling of the internal parts in a TENV and TEFC machine.
- Heat transfer through the bearings.

To characterize the manufacturing uncertainties, a DC current test is highly recommended and it is the most useful test to calibrate a thermal model. It is conducted by injecting a fixed DC current into the stator winding which generates a known copper loss in a machine while the rotor is static. To calibrate a thermal model effectively, all the phases are connected in series to give an equal current in all the slots so that the full machine can be heated up evenly. If not all the phases are energized, then a full machine is being heated irregularly. The winding temperatures could vary over a large range and sufficient thermocouples are required to capture the winding temperature variation for effective model calibration. To determine the copper loss, the voltage is measured directly across the windings to exclude the voltage drop across the leads. The total thermal loss can be then calculated by multiplying the voltage by the current.

The DC current test can be performed at different current levels but needs to ascertain the machine temperatures are below the temperature limit. Therefore, it is recommended to increase the DC current gradually, monitoring the temperature rise during the test. Figure 6.11 shows an example of how the coil temperature varies with different current input. For instance, a current brings the temperature to about 80 °C from ambient, then increases the current which will heat the coil to about 120 °C, then further increasing the current to further increase the coil temperature to around 160 °C. For each current applied, a reasonable time is given to reach steady state. After that, remove the heating and let the coil to cool down.

During the DC test, the machine temperature at various locations needs to be measured:

- Winding temperature, e.g. active winding temperature, coil temperature at slot opening, end winding temperature.

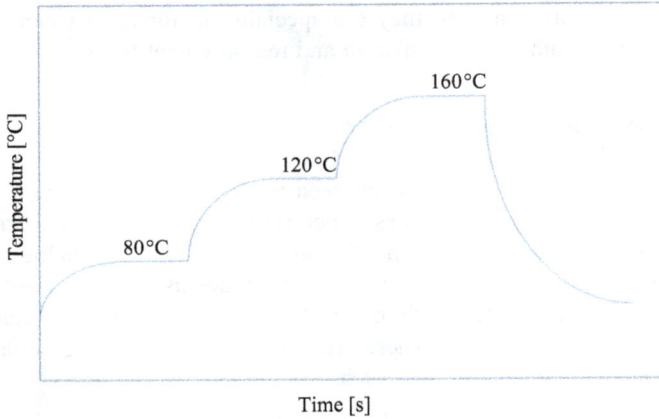

Figure 6.11   A schematic diagram shows coil temperature rise due the increase of input current

- Temperature of stator lamination temperature, e.g. stator lamination end surfaces, stator inner and outer diameter, stator tooth, stator yoke.
- Temperature of housing inner and outer surface temperatures.
- Temperature of the mounting plate, fixing, test bench, etc. as sometimes they can act as a heat sink.
- Thermocouples inside the end space to measure air temperature at NDE and DE. This will indicate the cooling/heating of rotor.
- Ambient temperature.
- Magnet and rotor temperatures can be measured using a thermal camera, Telemetry device, back emf, etc.

Based on the temperature measurements at various locations, if there are large temperature differences at certain nodes, this information can assist in identifying such issues as poor cooling of certain surfaces or through certain components due to manufacturing issues, additional cooling due to mounting (heat sink), etc. Often it is difficult to measure the winding hotspot as the location is unknown. However, the average winding temperature can be estimated easily by measuring the winding electrical resistance at ambient temperature and after steady-state using the formula below:

$$R_2 = R_1[1 + a(T_2 - T_1)] \tag{6.9}$$

where R_1 is the winding resistance at ambient temperature T_1, R_2 is the winding resistance at the average winding temperature T_2 after steady state. a is the resistivity coefficient of copper, which is 0.00393 K^{-1} at 20 C. The measured winding temperatures can then be compared with the average winding temperature for winding hotspot determination.

For accurate thermal model calibration, the test machine needs to be thermally insulated so that the DC copper loss is only dissipated to the main cooling system rather than to the mounting plate as a heat sink. Alternatively, if the machine is coupled to a large plate on the rig this test can be done with the machine on and off the rig. This enables calibration of any influence of the rig on cooling.

For the main cooling system, the heat dissipation by the cooling system (q) can be determined from the coolant temperature rise as:

$$q = \dot{m}c\Delta T \tag{6.10}$$

where $\dot{m}$ is the coolant mass flow rate, c is the specific heat, ΔT is the temperature difference between the inlet and the outlet. For liquid cooling, liquid coolants have high specific heat and therefore normally ΔT is only a few degree Celsius ($°C$). Consequently, it is recommended to use thermal sensors with high accuracy, or alternatively use more than one thermal sensor for inlet and outlet temperature measurements.

Essentially, a good correlation between simulation and reality is certainly achievable. However, in electrical machines, the challenges and uncertainties are typically in the manufacturing processes, e.g. electrical steel degradation, interface gaps, layout of conductors, etc. These uncertainties are often unknown at design stage. Through model validation exercises and understanding of manufacturing processes involved, high confidence in simulations can be achieved. The model validation is often also very useful in highlighting where issues and areas for improvement occur in the machine.

6.5 Thermal model calibration using a short transient test

A relatively new test is proposed [10,11] which gives focus to the heat transfer from windings to stator laminations. A short thermal transient test emulates an isothermal condition for the laminated stator core pack. This means that the stator winding can be considered an adiabatic system as the stator lamination temperature remains constant during the test. The winding-to-stator thermal resistance and winding thermal capacitance can both be accurately identified using this test.

The short transient uses stator DC excitation. The measured winding temperature data is curve fitted to give a first order system functional representation of the winding temperature which is used to find the winding to stator parameters. The calibration procedure involves initially defining the model regions with previously known or experimentally derived data. Later, the remaining unknown factors are estimated by tuning the model to match the model predictions with measured data.

The experimental procedure is suitable for both distributed and concentrated windings [10,12]. Ideally, the winding phases should be configured to give the same power loss in each slot with a series connection of all the phases. If this is not possible, the phases can be connected in parallel assuming that the current is equally shared among the phases.

Before starting the tests, the winding resistance at the ambient temperature (R_0) should be measured. The test machine should also be thermally insulated from the foot or flange mounted base.

An isothermal stator assembly can be assumed, if the averaged stator lamination temperature rise is less than 1 °C during the transient. For a steady-state test, a thermal equilibrium can be assumed when the winding temperature rise is less than 1 °C over a 30-min period.

A constant current is supplied to the winding. The winding resistance variation with time (R_T) can be monitored from measurements of the supply current (i_{dc}) and supply voltage (v_{dc}) using the relationship $R_T = v_{dc}/i_{dc}$. It is not required to have any thermocouples placed on the winding to measure the winding average temperature. This can be deduced from the equation:

$$T = \frac{R_T}{R_0}(B + T_0) - B \tag{6.11}$$

where B is the material temperature coefficient, a value equal to 234.5 K for copper. R_0 is the average winding resistance at the start of the test at an ambient temperature T_0.

During the transient test, the thermal energy stored can be calculated from the supply voltage and current measurements at each time:

$$W = v_{dc}(t)\, i_{dc}(t)\, t \tag{6.12}$$

A plot of W versus T is shown in Figure 6.12. The slope the straight-line energy versus temperature increase is the values of the winding (copper plus insulation) thermal capacitance in J/°C:

$$C_{wind} = \frac{dw}{dT} \tag{6.13}$$

The example shown in Figure 6.12 is for a test on a 4 kW induction motor. It has a slope or winding thermal capacitance of 1,759.9 J/°C.

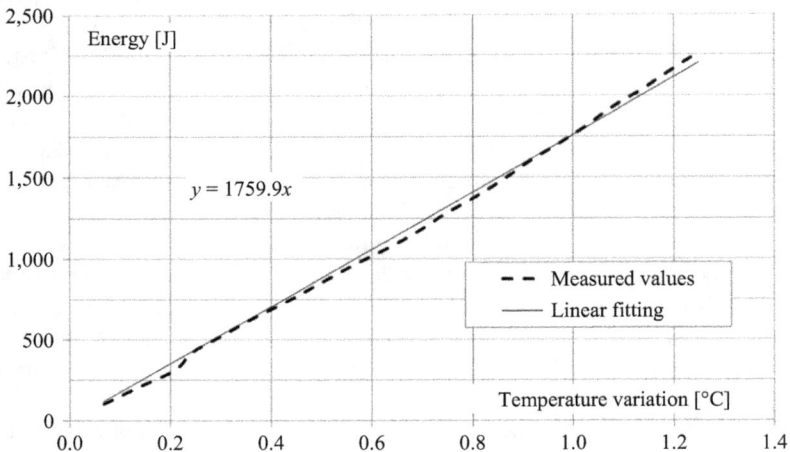

Figure 6.12   Winding thermal energy versus temperature variation

The computation of the thermal capacitances for components such as the stator lamination pack is quite straightforward as the component weight is easily calculated and the material specific heat is known. The winding is a bit more complex due to the combination of copper or aluminum and insulation materials. The test method described above can be useful to identify the winding total thermal capacitance.

The equivalent thermal resistance between the winding and the lamination $R_{cu\text{-}ir}$ can be determined from the mathematical representation of the first-order system defined by:

$$T_k = T_{k-1} + (T_0 + R_{cu\text{-}ir}P_{k-1} - T_{k-1})\left(1 - e^{-\frac{t_k - t_{k-1}}{R_{cu\text{-}ir}C_{wind}}}\right) \tag{6.14}$$

where P is the power loss and the indices k and 0 refer to time instant and initial condition, respectively. Such a procedure consists of minimizing the squared deviation between the winding temperature computations through (6.11) and (6.14). The optimization algorithm known as generalized reduced gradient (GRG) [13] can be used for this purpose.

An example of the results obtainable is shown in Figure 6.13. An excellent match is seen between the computed and the experimental prediction of average winding temperate versus time over the short thermal transient period.

The correct duration of the short transient test is difficult to identify but does depend on the motor size. Previous experience will aid in setting the test time. It is probably best to over-estimate the transient test time and neglect results where the winding can no longer be considered to be adiabatic, i.e. measured points not in agreement with a linear variation of the energy versus temperature variation as shown in Figure 6.14. In Figure 6.14, the winding can be considered in adiabatic

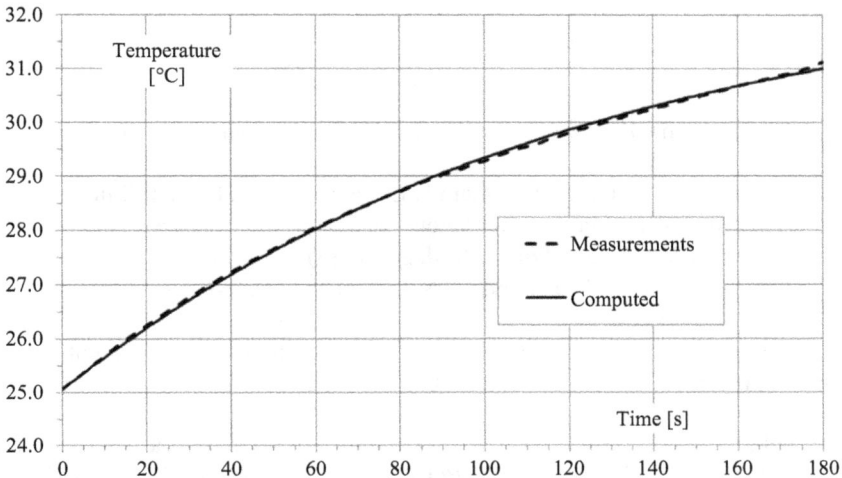

Figure 6.13   Winding temperature variation during short thermal transient

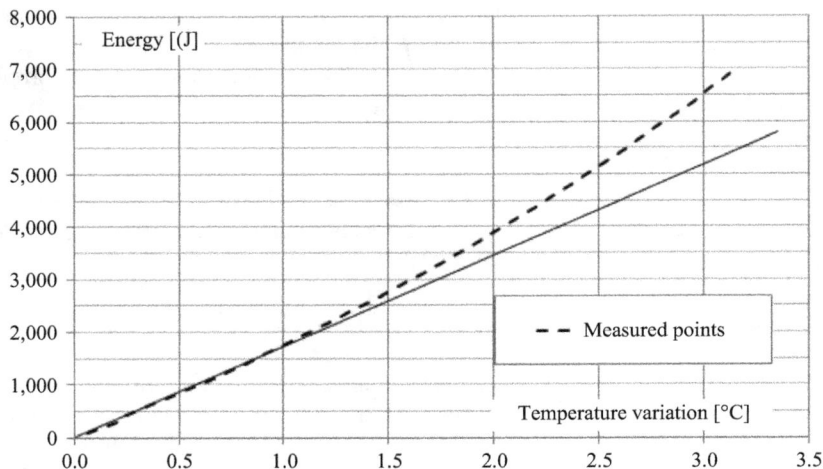

Figure 6.14   Winding thermal energy versus temperature variation for a "longer" short time transient

condition until the temperature variation is equal to 1.0–1.5 °C, where the energy trend is linear. For a temperature variation higher than 1.5 °C, the energy variation is no longer valid since the energy variation assumes a more "parabolic" trend. This means that other machine thermal capacitances come into play and the winding is not in an adiabatic condition. On the basis of the previous considerations, all the measured points corresponding to a temperature variation higher than 1.5 can be neglected.

References

[1] Bejan A. and Kraus A. *Handbook of Heat Transfer*. Hoboken, NJL John Wiley and Sons Inc.; 2003.
[2] Connor P.H., Pickering S.J., Gerada C., *et al.* 'Computational fluid dynamics modelling of an entire synchronous generator for improved thermal management', *IET Electr. Power Appl.* 2013;7(3):231–236.
[3] Tropea C., Yarin A. L., and Foss J. F. (eds.). *Handbook of Experimental Fluid Mechanics*. Berlin: Springer-Verlag; 2007.
[4] Simpson N., Wrobel R., and Mellor P. H. 'Estimation of equivalent thermal parameters of electrical windings', *IEEE Trans. Ind. Appl.* 2013;49(6): 2505–2515.
[5] Sullivan C. R. 'Optimal choice for number of strands in a litz-wire transformer winding', *IEEE Trans. Power Electron.* 1999; 14(2): 283–291.
[6] Wrobel R. and Mellor P. H. 'A general cuboidal element for three-dimensional thermal modelling', *IEEE Trans. Magn.* 2010; 46(8): 3197–3200.

[7]   Ayat S., Wrobel R., Goss J., and Drury D. 'Estimation of equivalent thermal conductivity for impregnated electrical windings formed from profiled rectangular conductors'. *8th IET International Conference on Power Electronics, Machines and Drives*, Glasgow, UK; 2016, pp. 1–6. doi: 10.1049/cp.2016.0313

[8]   Wrobel R., Simpson N., Mellor P. H., Goss J., and Staton D. A. 'Design of a brushless PM starter generator for low-cost manufacture and a high-aspect-ratio mechanical space envelope', *IEEE Trans. Ind. Appl.* 2017; 53(2): 1038–1048.

[9]   Cao W., Bradely K. J., and Ferrah A., 'Development of a high-precision calorimeter for measuring power loss in electrical machines', *IEEE Trans. Instr. Meas.* 2009; 58(3): 570–577.

[10] Boglietti A., Carpaneto E., Cossale M., *et al.* 'Stator winding thermal model conductivity evaluation: an industrial production assessment'. *2015 IEEE Energy Conversion Congress and Exposition (ECCE)*, Montreal, QC; 2015, pp. 4865–4871.

[11] Boglietti A., Carpaneto E., Cossale M., Staton D., Popescu M. 'Electrical machine first order short-time thermal transients model: Measurements and parameters evaluation', 40th Annual Conference of the IEEE Industrial Electronics Society (IECON), Dallas, USA, Nov 2014, pp. 555–561

[12] Boglietti A., Cossale M., Vaschetto S., and Dutra T. 'Thermal conductivity evaluation of fractional-slot concentrated-winding machines', *IEEE Trans. Ind. Appl.* 2017; 53(3): 2059–2065.

[13] Lasdon L. S., Waren A. D., Jain A., and Ratner M. 'Design and testing of a generalized reduced gradient code for nonlinear programming'. *J. ACM Trans. Math. Softw.* 1978; 4(1): 34–50.

Chapter 7

Application of design methods (case studies)

In this chapter, examples are given of practical electrical machine applications with different types of cooling such as Totally Enclosed Non-Ventilated (TENV), Totally Enclosed Fan-Cooled (TEFC), through ventilation, internal circulation flow with heat exchanger, housing water jackets, flooded stator cooling, and oil spray cooling. An indication is given of the reason that the cooling method was chosen for the particular application and some details of the advantages of one cooling type over another are explained. Apart from active cooling, examples are also given of electrical machine applications using more thermally conductive insulation materials. With enhanced passive cooling, the improvements in machine performance and torque/power density are explained.

Finally, an example is given of a high-performance electric motor designed to allow multiple cooling types to be used. The motor was for electric motorsport and the cooling can be tailored to the particular race circuit to give optimum overall performance.

7.1 Thermal management of electrical insulation system

In an electrical machine, the insulation system is chosen to meet the required design life. Electrical machines are usually designed with the maximum operating temperature below the thermal limit to give appropriate life, e.g. insulation life of 20,000 hours with the machine operating at full load. Based upon IEC standards, the common insulation classes B, F, and H have thermal limits of 130 °C, 155 °C, and 180 °C, respectively. The life of the insulation deteriorates rapidly if the electric motor is heated above the thermal limit. There is a rule of thumb for motor-insulation systems called the "10 °C half-life rule," i.e. every 10 °C increase in operating temperature cuts insulation life expectancy to a half. Increasing power and decreasing mass and space requirements see conventional power density limits being pushed with the thermal management of electrical machines becoming critical to ensuring reliability and robustness. Electrical machines continue to be pushed for higher operating voltages, elevated temperatures, and new cooling fluids. Therefore, the insulation materials must be able to maintain their electrical and mechanical properties in these aggressive environments. Besides good electrical properties, thermally conductive insulation materials would be advantageous to enhance the heat dissipation from the coils reducing machine temperature.

7.1.1   *Slot liner*

In this section, the impact of a more thermally conductive slot liner on the thermal and electromagnetic performance is analysed based on a benchmark – an early Nissan LEAF traction motor. The thermal model of the Nissan LEAF traction motor was built in Motor-CAD using machine data from tear down analysis. The Nissan LEAF traction motor is a 48 slot 8 pole interior permanent magnet (IPM) machine with distributed stranded windings. As the motor is mainly cooled by a liquid cooling jacket in the housing, radial heat conduction from the coils to the housing is the main heat flow path. Compared to the active parts (stator laminations and windings) which are mainly metallics, the insulation materials have much lower thermal conductivities. They cause considerable thermal resistance restricting heat flow from Joule heating in the coils to the housing. In the study, the slot liner of the model is a 0.25 mm thick NMN 3-3-3 laminate with an effective thermal conductivity of 0.14 W/m/K. Figure 7.1 shows the impact of slot liner thermal conductivity on winding temperature at steady state based on the operating point: 183 Nm at 4,000 rpm. The sensitivity analysis demonstrates that the maximum winding temperature can be reduced by up to 13.6 °C with slot liner thermal conductivity increasing from 0.12 W/m/K to 0.8 W/m/K.

Typically, the thermally limited characteristics of an electrical machine over the full speed range is represented by the continuous performance curve. The continuous performance characteristics are calculated by setting maximum temperature limits for different components of the machine and calculating the maximum torque that can be achieved continuously within those limits across the full speed range. This requires the solution of the electromagnetic, loss, and thermal models simultaneously. Machine continuous performance is commonly limited by

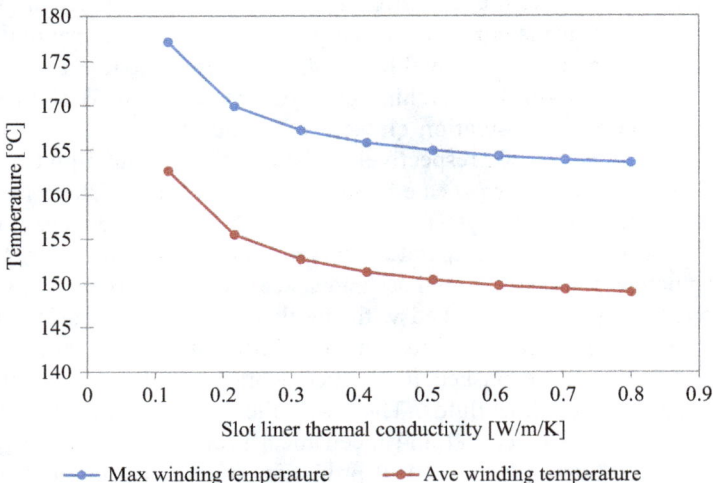

Figure 7.1   The impact of slot liner thermal conductivity on winding temperatures

the winding temperature at low speed, as the copper loss dominates, while at high speed, the continuous torque is more limited by the magnet temperature due to induced eddy currents. Consequently, thermally conductive slot liner materials can improve the machine continuous performance particularly at low speed. With thermal limits of 180 °C for the stator winding based on insulation class H and 140 °C for the magnet, Figure 7.2 illustrates the continuous torque performance increases with higher slot liner thermal conductivity especially when the speed is less than 4,000 rpm. Table 7.1 shows the improvement of continuous torque at zero rpm (stall torque) for different slot liner thermal conductivities compared to the baseline of slot liner thermal conductivity of 0.12 W/m/K. The proposed solution demonstrates that the stall torque can be increased by up to 6%.

Alternatively, a more thermally conductive slot liner material leads to lower copper loss as the winding electrical resistance is strongly dependent on winding

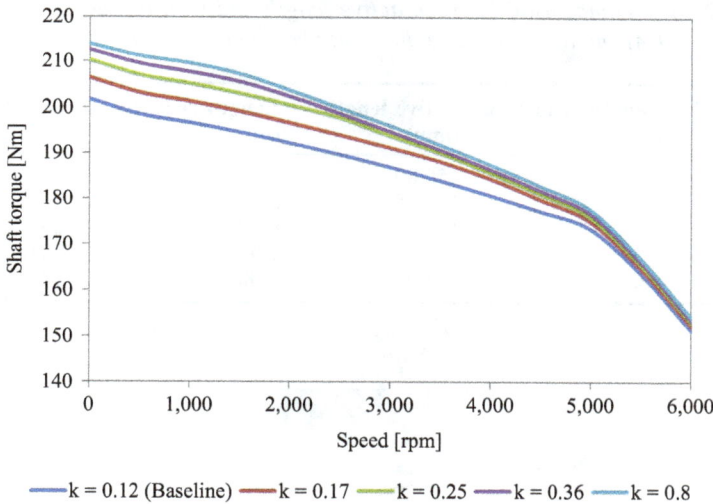

Figure 7.2   The impact of slot liner thermal conductivity on continuous torque performance at low speed region

Table 7.1   Improvement of stall torque with more thermally conductive slot liner

Slot liner k (W/m/K)	Stall torque (Nm)	Increment (Nm)	Percentage
0.12	201.7	–	–
0.17	206.4	4.7	2.3
0.25	210.2	8.5	4.2
0.36	212.5	10.8	5.3
0.80	213.9	12.2	6.0

temperature as described in Section 4.1.7.1. Hence, machine efficiency can be improved with a more thermally conductive slot liner material.

A higher thermal conductivity slot liner enables higher current for the same thermal limit. This could potentially save material cost by redesigning the Nissan LEAF motor with a shorter active length. For permanent magnet machines, the shaft torque is directly proportional to the active length. The percentage improvement in continuous torque as shown in Table 7.1, while the reduction in active length is given in Table 7.2. Similar stall torque can be achieved with a shorter active length by using a higher thermal conductivity slot liner material. By using a slot liner with a thermal conductivity of 0.8 W/m/K compared to the baseline of 0.12 W/m/K, it shows that the machine active length can be reduced up to 6 % which is equivalent to 9 mm saving. Figure 7.3 shows the continuous performance

Table 7.2   *The comparison between active length and stall torque of the Nissan LEAF motor with more thermally conductive slot liner*

Slot liner k (W/m/K)	Length reduction (%)	Active length (mm)	Length reduction (mm)	Stall torque (Nm)
0.12	–	150.00	–	201.7
0.17	2.3	146.55	3.45	201.6
0.25	4.2	143.70	6.30	201.2
0.36	5.3	142.05	7.95	201.4
0.80	6.0	141.00	9.00	201.5

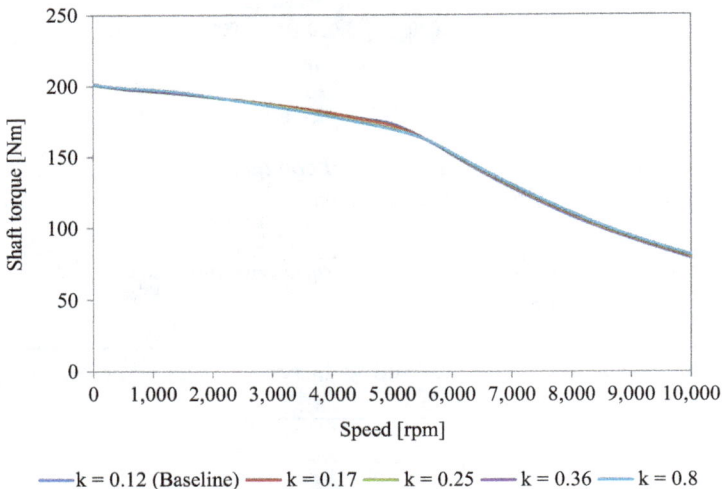

—— k = 0.12 (Baseline) —— k = 0.17 —— k = 0.25 —— k = 0.36 —— k = 0.8

Figure 7.3   *The continuous performance characteristics of different slot liner thermal conductivities simulated with shorter active length*

Table 7.3   The reduction in active material weight with more
thermally conductive slot liner

Slot liner k (W/m/K)	Active materials weight (kg)	Weight reduction (kg)	Weight reduction (%)
0.12	31.80	–	–
0.17	31.12	0.69	2.2
0.25	30.55	1.25	4.1
0.36	30.22	1.58	5.2
0.80	30.01	1.79	6.0

characteristics of different slot liner thermal conductivities simulated with shorter active length while maintaining the same thermal limits. Due to the reduction in active length, motor redesign enables a reduction in the quantity of electrical steel, magnet and copper, giving up to 1.79 kg active material savings as given in Table 7.3.

7.1.2   Impregnation resin

In this section, the impact of thermally conductive impregnation resins on thermal and electromagnetic performance is analysed based on a benchmark – 2016 BMW i3 traction motor. The electric machine designed by BMW shows the latest state of development for IPM machines with a high reluctance torque to reduce the cost of the rare earth magnets. The motor is a 72 slot 12 pole IPM machine with double layer magnet configuration. This configuration is used to minimize the permeability of the *d*-axis. This results in a difference in inductance between the magnet axis (*d*-axis) and the interpolar axis (*q*-axis), maximizing the potential to utilize reluctance torque. The simulation model of the BMW i3 motor is built in Motor-CAD based on BMW published data and machine data from a tear down study. The machine is mainly cooled by a liquid cooling jacket in the housing similar to Nissan LEAF motor in Section 7.1.1. As well as the slot liner, the impregnation resin used in the slot also has low thermal conductivity, typically 0.2 W/m/K. This restricts the conduction heat flow from the copper conductors to the housing.

Based on operating point of 170 Nm at 4,500 rpm, a sensitivity analysis was performed by varying the thermal conductivity of the impregnation material from 0.2 to 1 W/m/K to investigate the impact on winding temperature rise at steady state. In the analysis, the other thermal parameters remain the same. As shown in Figure. 7.4, the maximum hot-spot winding temperature reduces from 166.1 °C to 157 °C, a temperature reduction of 9 °C. This indicates that with the proposed increased resin thermal conductivity, a higher stator current could be allowed within the insulation thermal limit.

Figure 7.5 illustrates the continuous performance curve of BMW i3 motor with impregnation resin thermal conductivity varying from a baseline of 0.2 to 1.0 W/m/K. To model this, the machine electromagnetic, loss, and thermal models are solved simultaneously. The maximum stator winding temperature is limited to 180 °C

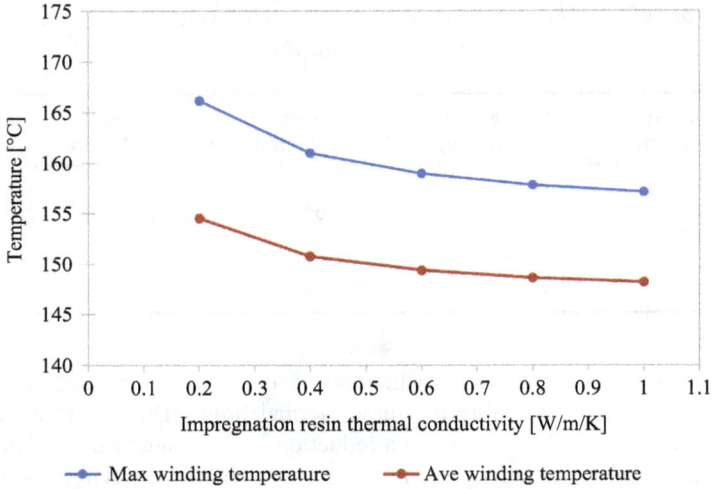

Figure 7.4   The impact of impregnation resin thermal conductivity on winding temperatures

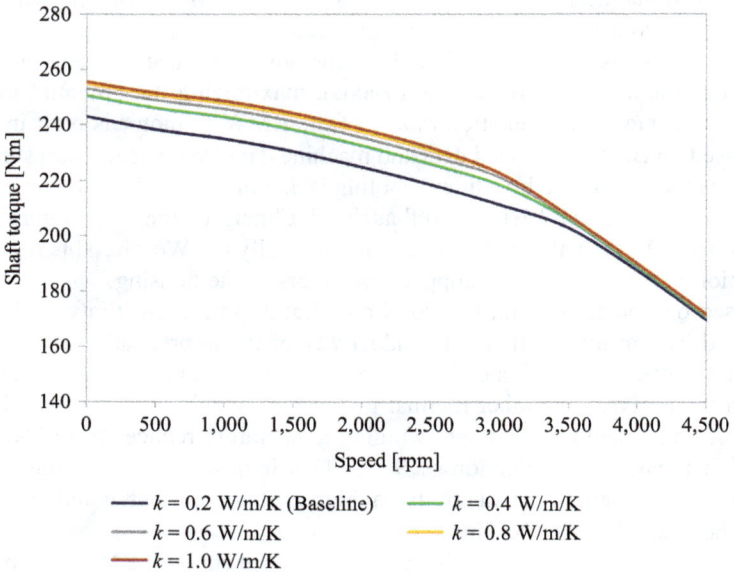

Figure 7.5   The impact of impregnation resin thermal conductivity on continuous torque performance at low-speed region

(insulation class H), while magnet temperature is limited to 150 °C. As shown in Figure. 7.5, the continuous torque increases with a higher thermal conductivity impregnation resin especially in the low-speed region (i.e. less than 3,000 rpm) as the continuous torque is more restricted by winding temperature at low speed. Whereas at high speed, the continuous torque is more restricted by magnet temperature. It is important to note that the impact of impregnation resin thermal conductivity on magnet temperature is not very significant. Table 7.4 shows the amount of improvement for the continuous torque at zero rpm (stall torque) for different impregnation resin thermal conductivities compared to the baseline of 0.2 W/m/K. The proposed solution demonstrates that the stall torque can be increased by up to 5.3%.

Alternatively, due to the lower temperatures, a higher stator current could be allowed within the thermal limits. The use of a more thermally conductive impregnation resin could potentially save material cost. The redesign of BMW i3 motor can be considered by calculating the reduction in active length that gives the same continuous torque performance as the case with the lower impregnation resin thermal conductivity of 0.2 W/m/K. For permanent magnet machines, the shaft torque is directly proportional to the active length. Hence, the percentage improvement in continuous torque leads to a percentage reduction in active length. The simulation results are given in Table 7.5, similar stall torques can be achieved

Table 7.4   Improvement of stall torque with more thermally conductive impregnation resin

Impregnation resin k (W/m/K)	Stall torque (Nm)	Increment (Nm)	Percentage
0.2	242.7	–	–
0.4	250.3	7.6	3.1
0.6	253.0	10.3	4.2
0.8	254.5	11.8	4.9
1.0	255.4	12.7	5.3

Table 7.5   The comparison between active length and stall torque of the BMW i3 motor with more thermally conductive impregnation resin

Impregnation resin k (W/m/K)	Length reduction (%)	Active length (mm)	Length reduction (mm)	Stall torque (Nm)
0.2	–	130	–	242.7
0.4	3.13	125.9	4.1	241.8
0.6	4.24	124.5	5.5	241.6
0.8	4.85	123.7	6.3	241.3
1.0	5.25	123.2	6.8	241.2

with a shorter machine active length by using higher thermal conductivity impregnation resin. Using an impregnation resin with a thermal conductivity of 1.0 W/m/K compared to the baseline, shows that the machine active length can be reduced up to 5.25%, equivalent to a 6.8 mm reduction in active length. The comparison between the continuous torque curves for different resin thermal conductivities across the full speed range is depicted in Figure. 7.6. Due to the reduction in machine active length, the motor redesign enables a reduction in the quantity of electrical steel, magnet and copper, giving up to 1.59 kg savings in active material mass as shown in Table 7.6.

Insulation materials typically have relatively low thermal conductivity restricting heat dissipation from the winding. Improving the passive thermal design through more thermally conductive insulation materials can reduce the winding temperature rise and also increase machine efficiency. This section has shown the

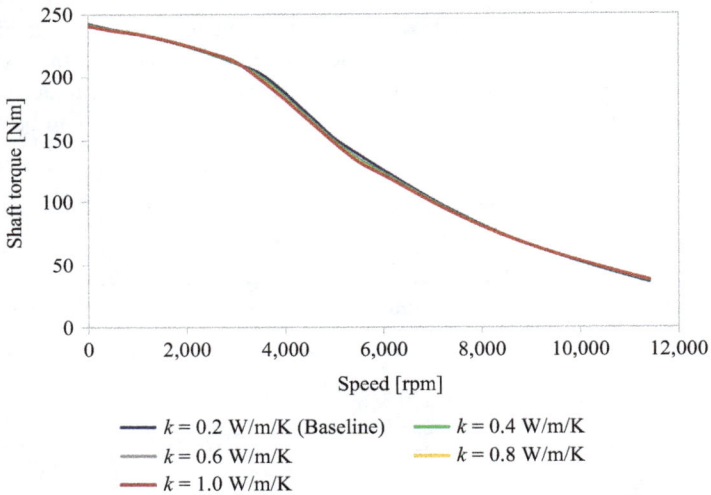

Figure 7.6   *The continuous performance characteristics of different impregnation resin thermal conductivities simulated with shorter active length*

Table 7.6   *The reduction in active material weight with more thermally conductive impregnation resin*

Impregnation resin k (W/m/K)	Active material weight (kg)	Weight reduction (kg)	Weight reduction (%)
0.2	34.39	–	–
0.4	33.43	0.96	2.79
0.6	33.1	1.29	3.75
0.8	32.91	1.48	4.30
1.0	32.8	1.59	4.62

potential for using more thermally conductive insulation materials to improve the torque/power density of an electrical machine by reducing the component size and weight while maintaining performance within the same thermal limits.

It is important to note that the impregnation material of thermal conductivity of 0.2 W/m/K is typically an unfilled epoxy resin. To increase the value of thermal conductivity, thermally conductive filler can be added to the impregnation system which can facilitate heat transfer. Impregnations and encapsulation resins supplied by Huntsman [1,2] can get to 1.2 W/m/K while compromising electrical, chemical and mechanical properties. CoolTherm® thermal management epoxy resins provided by LORD are formulated for encapsulation of motor windings combined with high thermal conductivity and excellent electrical insulation. For instance, CoolTherm EP-340-filled epoxy resin has thermal conductivity of 1.4 W/m/K while providing outstanding mechanical properties with a very low coefficient of thermal expansion [3].

7.2   Totally Enclosed Non-Ventilated cooling

In Section 4.2, the addition of radial fins to a motor housing to improve the natural convection cooling of a Totally Enclosed Non-Ventilated (TENV) motor has been addressed. The problem is that radial fins only give a benefit if it can be guaranteed that motor has a horizontal shaft orientation. If the motor can be mounted with any orientation, then a special angled fin design like that shown in Figure 7.7 could be used. This allows roughly equal natural airflow between fins with any orientation.

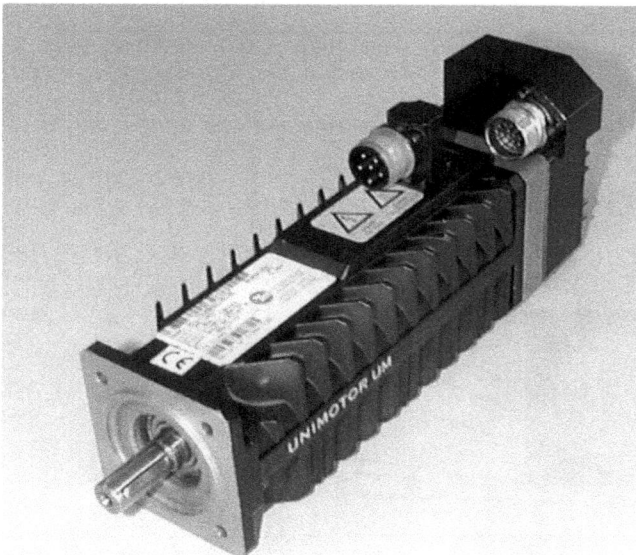

Figure 7.7   TENV motor with angled fin design

A practical example of a servo motor running from a duty cycle load is shown in Figure 7.8 [4], which shows the duty cycle thermal performance of a servo motor predicted using the lumped parameter thermal network created by Motor-CAD and compared to test data. The repetitive duty cycle is a short period at three times the rated current followed by a longer period at half rated current. The motor winding heats up rapidly during the overload periods and cools down during the light load condition. The bulk of the motor heats up at a much slower rate and reaches a steady state after around 200 min or around 20 cycles. The maximum winding temperature then rises from around 95 °C to 150 °C during the overload periods.

Figure 7.9 shows some work done by Goodrich Power Systems on the optimization of a short duty cycle rated motor [5]. Motor-CAD was used to

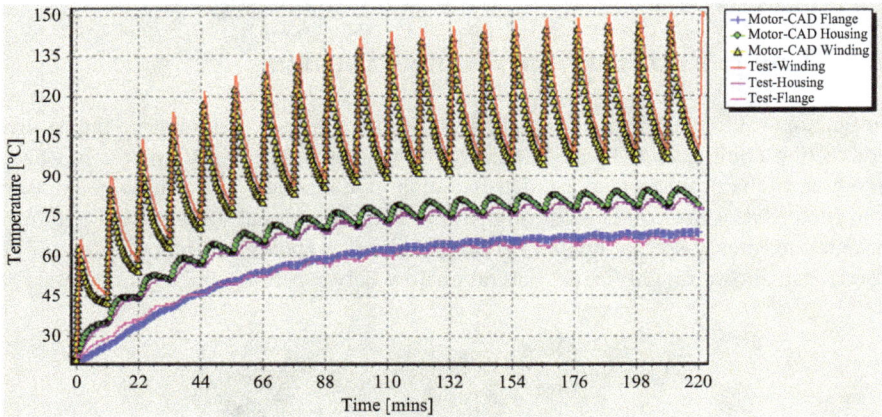

Figure 7.8   Thermal transient of a servo motor operating a duty cycle type load (test data also shown)

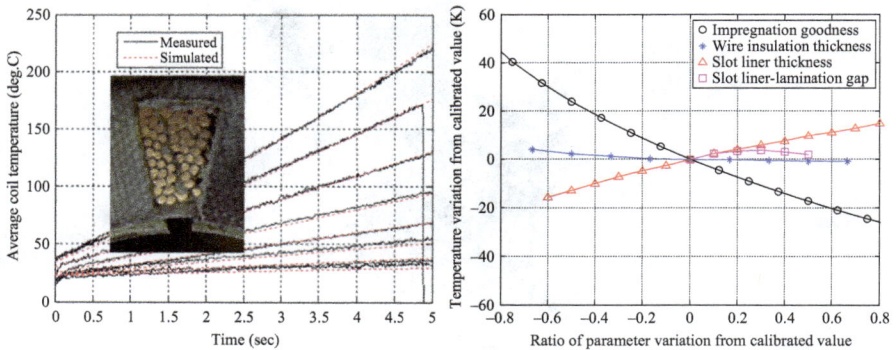

Figure 7.9   Thermal modelling of a short-duty motor using Motor-CAD

optimize the quality of the slot impregnation as it was found to play a significant role in the transient performance. Optimization was facilitated by the in-built multi-parametric sensitivity analysis module in Motor-CAD, Figure 7.9 showing the results of sensitivity analysis on various slot insulation parameters, i.e. how well the winding is impregnated, the wire enamel thickness, the slot liner thickness and the thickness of the gap between the slot liner and the lamination. In this case, the impregnation goodness (amount of air left in the impregnation system) was found to be the most important factor and the motor winding was optimized to account for this. Thermal analysis also showed that thermal protection using temperature sensors was not reliable due to a delayed response in temperature measurement and that a predictive protection was essential. The analysis was confirmed by 2D finite element analysis and prototype motor testing, excellent agreement with measured thermal transients for a range of loads being shown in Figure 7.9.

7.3 Totally Enclosed Fan-Cooling

Figure 7.10 shows a typical industrial range of Totally Enclosed Fan-Cooled (TEFC) induction motors with rated powers of 4, 7.5, 15, 30, and 55 kW [6,7]. This range of motors has been extensively modelled and tested to give a thorough understanding of TEFC cooling. In order to estimate the convection heat transfer coefficient within the fin channels, the local air velocity needs to be estimated. This is complicated by the fact that different channels often have different local velocities due to rotational flow effects and blockage of the air flow by fan cowling supports, terminal boxes, etc. Figure 7.11 shows measured local flow velocity values in different fin channels in the 4 kW motor. Measurements are shown for three positions along the axial length of the frame (rear/fan end, centre, front/shaft end).

Figure 7.10   Typical range of TEFC induction motors [6,7]

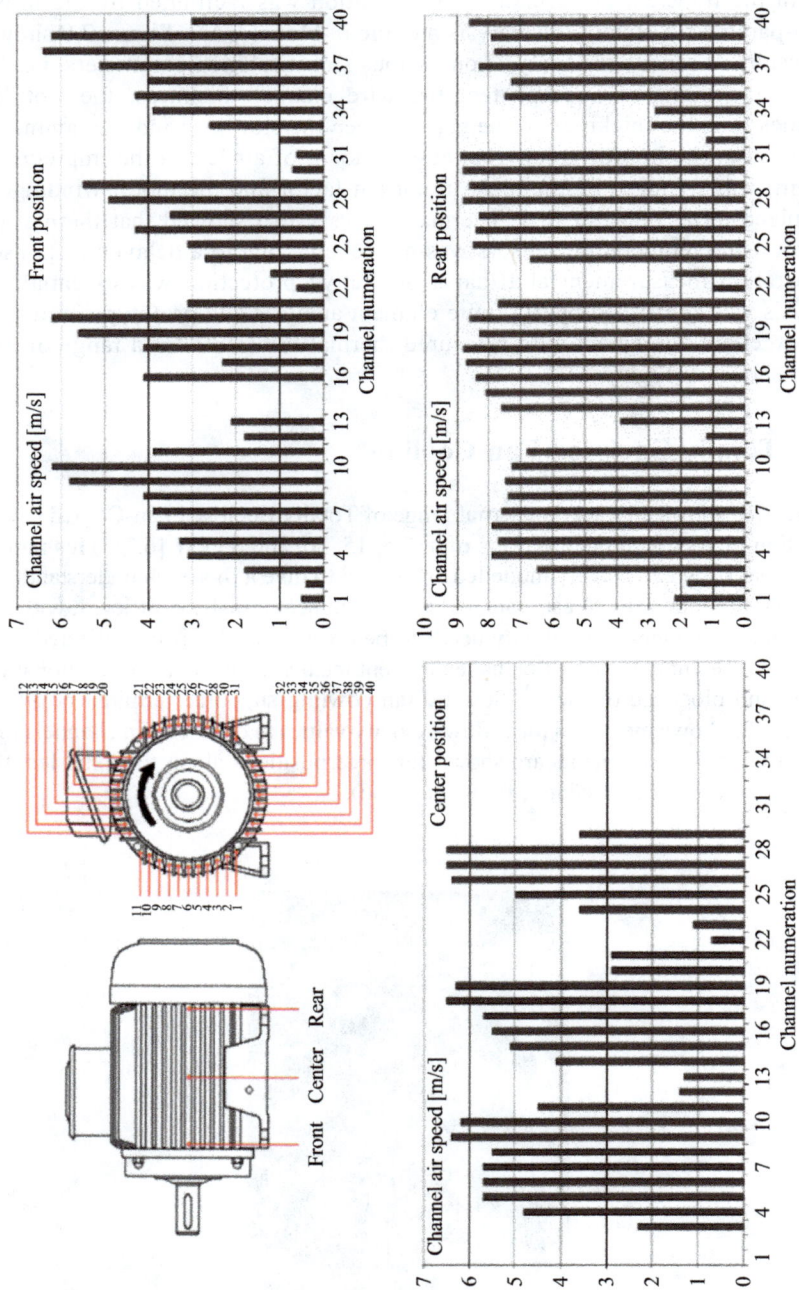

Figure 7.11 Typical form of variation in fin channel air velocity with fin position and along the axial length, 4 kW motor [6,7]

The computed average air speed values in the rear, centre and front positions are reported in Table 7.7 [7]. In the rear position (fan cowling outlet), the air speed is double with respect to the front position. This is in line with the axial open fin channel leakage graphs shown in Figure 4.27.

Figure 7.12 shows the typical accuracy that can be expected with an un-calibrated model in Motor-CAD. Here the default parameters in Motor-CAD have been used and the effective thermal resistance between housing and ambient calculated. Typical open channel air leakage data is used in this case (default values in Motor-CAD which are the average of characteristic shown in Figure 4.27 and reference [6]). It is seen that an accurate estimation can be made if the user has a basic knowledge of the inlet air velocity or volume flow rate to the fin channels. From Figure 7.12, it is seen that the larger machines tend to have a higher air speed, this being confirmed by the measured housing air velocity data for different sized machines shown in Figure 7.13. Further analysis of the data shown in Figure 7.13 shows that for similar designs of fan the air velocity is roughly directly proportional to the fan diameter at a particular speed.

Table 7.7   *Average air speed axially along the housing fin channels*

Position	Rear	Center	Front
Air speed (m/s)	6.7	4.2	3.2

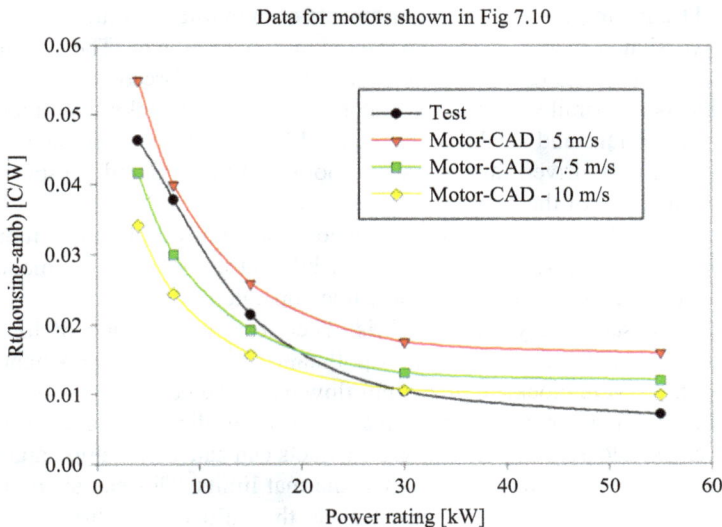

Figure 7.12   *Forced convection resistance between frame and ambient air [6]*

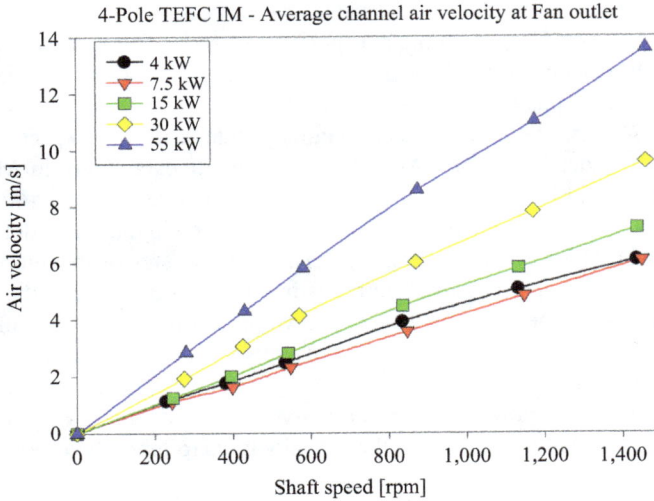

Figure 7.13   Variation in fin channel air velocity (average at fan outlet) with rotational speed for the motors in Figure 7.10 [6]

7.4   Open ventilated cooling

Open ventilated cooling is one of the common cooling methods used by electric traction motors in railway applications (tram, metro, light rail vehicle, locomotive, high speed train, etc.) to meet the requirements of high torque/volume ratio [8]. For open self-ventilated motors, a fan is mounted on the rotor shaft. The external air is drawn into the motor to remove the heat generated. Open forced ventilated cooling is typically used by railway traction motors to ensure the highest thermal performance. External fans and air ducts are required for open forced ventilated cooling. By using externally driven fans, constant cooling can be assured irrespectively of the rotational speed of the motor.

Due to the demands of low life cycle costs, railway traction motors are required to run over a range of millions of kilometres between the maintenance intervals. However, the motors are sometimes operated in a polluted environment involving dust, sand, salty air, or debris. Therefore, a challenge of long maintenance intervals is the accumulation of pollutants in electrical components which can lead to the severe blockage of cooling flow paths, especially when the light rail vehicles operate in an environment with a high level of dirt, e.g. trams operating in suburban areas. Blockage of the cooling channels can cause the temperature to rise in the windings and bearings beyond their thermal limits. This can shorten the life of the insulation system and bearings. Hence, the influence of dirt accumulation needs to be studied during the design phase, in order to address any potential failure during service operation.

For railway traction motors that involve transient running profiles with lots of acceleration and deceleration, the lumped parameter thermal network is the most suitable method because of its computational effectiveness. However, the equivalent thermal network of a traction motor does not usually have the level of detail to allow the individual cooling channels to be modelled separately, especially when the blockage could be random leading to asymmetric cooling. Consequently, a coupled finite-element and lumped parameter (FE–LP) modelling method was proposed in Ref. [8]. As shown in Figure 7.14, the active section comprising cooling channels is modelled by a 3D finite-element model while the other motor parts are modelled using lumped parameter thermal network. CFD is also employed to define the convective boundary condition for the finite element model based upon the flow distribution in the motor between the stator ducts, air gap and rotor ducts. Hence, the thermal analysis of stator and rotor duct blockage can be performed using the proposed modelling technique for accurate estimation of motor thermal behavior in real operation and also motor life time prediction.

The motor used for this study is an induction motor with rated voltage and power of 750 V and 250 kW at 2,500 rpm. The motor investigated has 16 and 19 cooling ducts in the rotor and stator, respectively. The diameter of the rotor ducts is 17 mm and they are located at a radius of 37 mm from the shaft. Stator ducts are located between the stator and housing with a width and height of 41 mm and 10 mm, respectively. Temperature measurements in the motor over a specified duty cycle were performed to validate the proposed FE-LP model, as depicted in Figure 7.15.

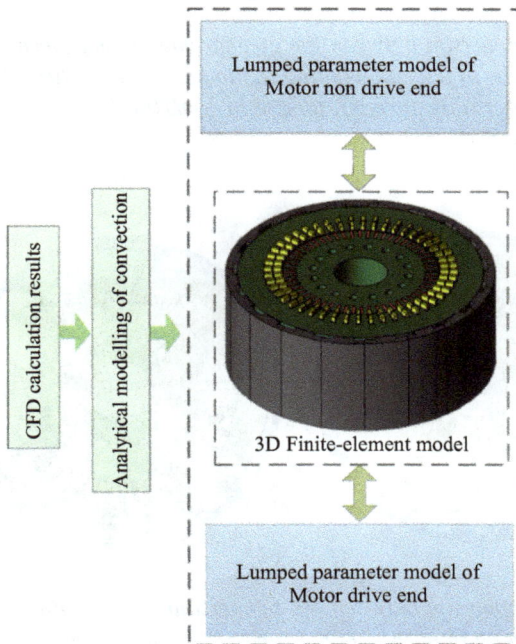

Figure 7.14   A coupled FE–LP modelling method [8]

The validated FE-LP model is further used to analyse the temperature rise of traction motor with certain cooling ducts blocked. Figure 7.16 shows the comparison between the temperature distribution in the motor running under normal operation and a motor whose four stator ducts at the bottom of the motor are blocked. The blockage results in higher temperature at the lower part of the motor. Moreover, to investigate the impact of rotor duct blockage on bearing temperature rise, two calculations were performed. One calculation was for the motor running under normal condition and the other calculation was for the motor with rotor ducts blocked. The thermal simulations given in Figure 7.17 indicate the risk of bearings being overheated when the rotor ducts are blocked. Therefore, it is critical to have an advanced thermal modelling method that allows machine designers to analyse the thermal behavior of railway traction motors over

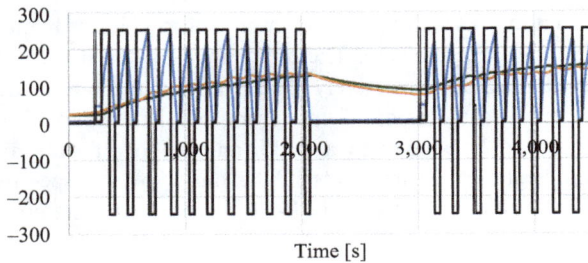

Figure 7.15 Comparison between the measurements and thermal modelling results of the end winding. Estimated temperature (brown), measured temperature (green), torque in Nm/10 (black), and speed in rpm/20 (blue) [8]

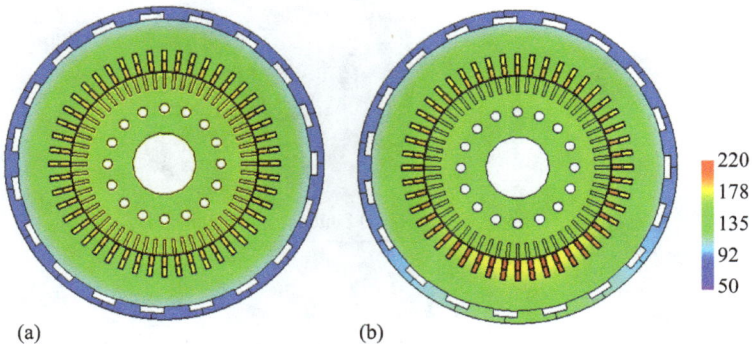

Figure 7.16 Comparison between the temperature distributions in the (a) motor running in normal condition and (b) motor whose four stator ducts at the bottom of the motor are blocked [8]

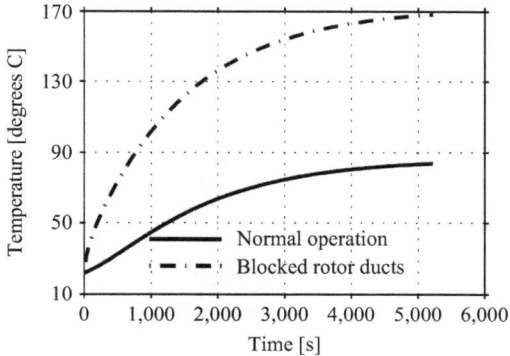

Figure 7.17 Comparison of the bearing temperature rise for the motor running under normal condition and the other calculation is for the motor with rotor ducts are blocked [8]

complex transient running profiles. Furthermore, the modelling method is also capable of modelling individual cooling ducts of an open ventilated electric motor with the risk of ducts blockage.

For an open ventilated machine, the air flow rate passing through the machine and the flow distribution between the cooling paths is critical in affecting thermal performance. However, the air flow rate cannot be solved directly because the flow rate through the flow system is determined from the intersection between the fan characteristic and system flow resistance curves as explained in Section 3.2. Furthermore, the system flow resistance is affected by the shaft speed as the effects of rotation can potentially change the flow distribution between the cooling channels due to the additional rotational pressure loss. CFD is a common modelling method used but the solution time is too long, especially when modelling a machine that is operated under transient operation. As an alternative to CFD, a coupled flow network and thermal network analysis method is an attractive solution because of the fast calculation speed that allows design engineers to perform what-if analysis. The use of flow network analysis to determine the required air flow rate to cool the stator winding hot spot below 160 °C was demonstrated by Ref. [9] for an 850 kW vector controlled induction motor for oil field drilling applications. Figure 7.18 shows the intersection of the fan characteristic and system flow resistance characteristic for the motor. The system flow resistance depends upon the duct wall friction and changes in flow condition such as contractions and expansions in the flow circuit and restrictions due to obstructions in the flow path. Then, local convective heat transfer in the cooling channels can be calculated from the local air velocity. The predicted temperatures were found to have good agreement with the measurements, i.e. the measured winding hot spot being 157 °C and the predicted value being 161 °C when the motor is cooled by 3,300 ft^3/min of air flow from the externally driven fan.

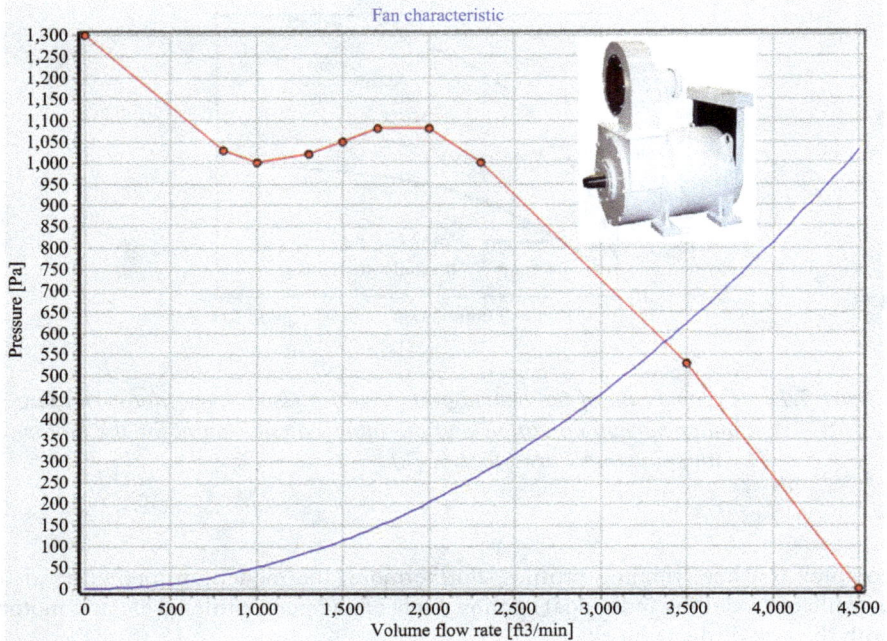

Figure 7.18   The fan characteristic and system flow resistance of 850 kW
induction motor

7.5   Close circuit cooling with a heat exchanger

Close circuit cooling is commonly used by high-voltage machines, e.g. P-series
Laurence Scott machine [10]. The induction machines can generate output power
up to 20 MW for supply voltages from 3.15 kV to 13.8 kV at 50 Hz or 60 Hz.

Pre-formed rectangular copper conductors are used for the stator winding and
each coil is insulated with mica tape. For high-voltage machines, the slot portion
of the coil is insulated with a resistive corona shield to reduce voltage stress on the
edges of core laminations. The coils are assembled into the open slots of the stator
core pack. The whole wound stator assembly is then consolidated under vacuum
pressure impregnation (VPI). This is to ensure maximum resin penetration to
enhance winding heat transfer. To effectively dissipate the heat generated in the
stator core pack, there are some radial ventilating ducts spaced uniformly along the
stator core. This will allow cooling air to have direct contact with the heat sources.
The rotor cage consists of copper bars. The rotor core pack is shrunk fit onto a
spoke-type rotor shaft. The spoke-type rotor allows the cooling air to pass into the
rotor axially by means of shaft-mounted fans. Similar to the stator, the rotor is
cooled by the radial ventilating ducts spaced at the same intervals along the
rotor core.

Both the stator and the rotor are supported by a structural frame inside a fully sealed enclosure to meet the mechanical protection, e.g. IP55, IP56. Either an air or water cooled heat exchanger is mounted at the top of the enclosure. The hot exhaust air from the radial ventilating ducts is cooled by the heat exchanger before being returned to the stator and rotor due to the differential pressure induced by the shaft-mounted fans. The box frame structure machines have been proven to provide reliable and cost effective operation in a wide variety of industries and environments. Also, the machines can be used in many arduous and difficult environments, including the largest off-shore induction motor currently in operation on a North Sea Platform rated at 14.5 MW and another in the Caspian area rated at 15.3 MW.

Compared to air-cooled turbo generators, hydrogen-cooled turbo generators provide much higher output power in a smaller footprint due to the reduced weight of the active components. In addition, hydrogen-cooled generators have a higher efficiency than air cooled units due to lower ventilation losses. Also, insulation life is extended due to the higher performance of the cooling medium. The use of closed circuit cooling with hydrogen as the cooling medium can be found in Ref. [11]. The Brush combined hydrogen and water cooled 2-pole turbo generators have output ranges from 250 to 1100 MVA with voltage ranges from 15 to 24 kV. BRUSH combined cooled generators are extensively used in steam turbine based power generation applications, e.g. Temelin Power Station in the Czech Republic. The stator winding is cooled directly by chemically pure water and the stator winding water circuit is supplied and discharged by PTFE (Teflon) hoses with a centrifugal pump. On the other hand, the rest of the machine is cooled directly with pressurized hydrogen. Hydrogen is forced around the generator by means of two axial flow fans mounted on the rotor shaft. Cooling circuits are designed to cool the windings as uniformly as possible. The hot exhaust hydrogen is cooled by hydrogen/water heat exchangers before being returned to the inlet. The removal of losses is relatively simple but the combined hydrogen and water cooling is a very efficient process that ensures maximum utilization of active materials.

Besides turbo generators, the use of closed circuit cooling can also be found for electric traction motors. For permanent magnet synchronous machine (PMSM), the power output is limited by its thermal performance as the magnet is very sensitive to temperature – elevated temperature can lead to irreversible demagnetization of permanent magnet. Therefore, active cooling is required for the rotor. As shown in Figure 7.19, air is circulated between cooling channels in the rotor and in the housing by a shaft-mounted fan; the air is cooled as it passes through the housing acting as a heat exchanger between the air and the water jacket. When compared to housing water jacket cooled machines, the advantage of this system with closed circuit cooling is to reduce the air temperature inside the machine, hence providing better cooling to the rotor and magnets while the motor is still a sealed, closed unit. As the maximum operating temperature of the magnets is lowered, this potentially enables a cheaper grade magnet to be used with less dysprosium. The use of this cooling configuration can be demonstrated by Zytek 170 kW 460 Nm electric traction motor [12].

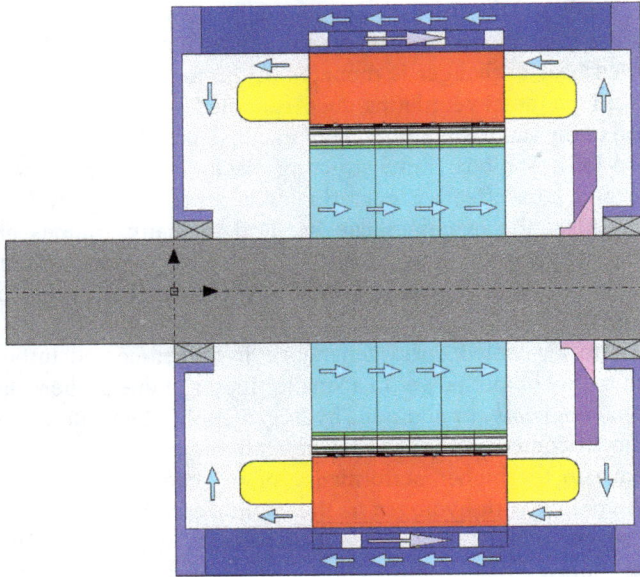

Figure 7.19   A permanent magnet synchronous motor with closed circuit cooling

7.6   Housing water jacket cooling

For hybrid and electric vehicles, the electric motors need to generate sufficient torque to propel the vehicles. However, the space that is available for the motors is often very limited. Therefore, this results in electric traction motors with high torque density and hence an effective cooling system is required. By reviewing the automotive traction motors, their main cooling system is a liquid cooling jacket in the housing due to its high cooling performance as described in section 4.6 while meeting the ingress protection requirement.

Typically, the coolant used is an ethylene glycol and water solution (EGW) which it is already available in the vehicle cooling system. Figure 7.20 illustrates the axial geometry of Nissan LEAF traction motor [13,14]. The Nissan LEAF motor is an IPM synchronous machine. Its housing cooling jacket consists of three circumferential channels that are connected in series. The cross-sectional area of the circumferential channels is relatively large to reduce the flow resistance to meet the pumping pressure.

Figure 7.21 illustrates the axial geometry of a Tesla Model S traction motor with its cooling system. The Tesla Model S traction motor is an induction machine with a copper rotor cage. The channel grooves are created in the inner housing and the housing cooling jacket is formed by welding the inner and outer housings together [15]. Compared to the Nissan LEAF motor in Figure 7.20, the cross section of each channel is relatively small. To reduce the pressure drop, there are 6 channels connected in parallel. The design can potentially provide uniform cooling in

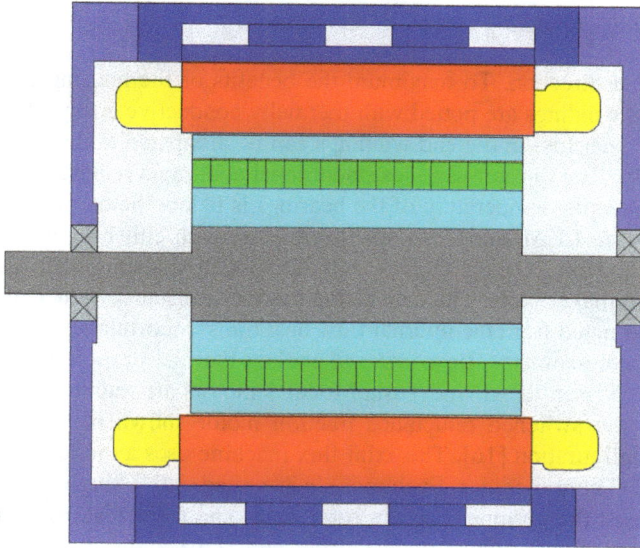

Figure 7.20 Nissan Leaf traction motor and its housing cooling jacket design

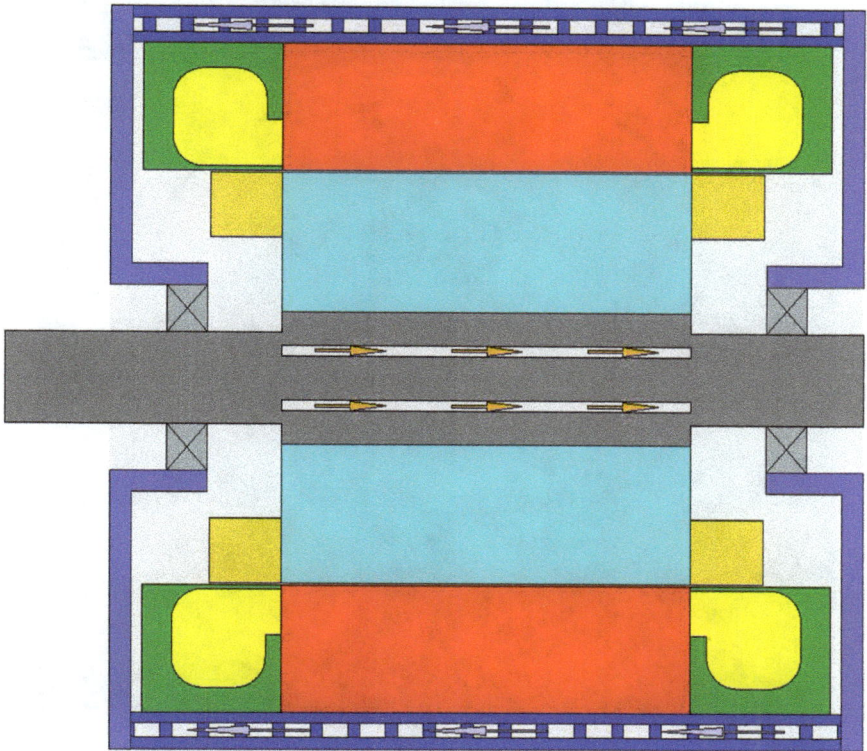

Figure 7.21 Tesla Model S traction motor with housing cooling jacket and shaft cooling

circumferential direction. To maximum the benefits of the housing cooling jacket, the stator end windings are potted with thermally conductive encapsulation material so that the copper loss in the end windings can be dissipated to the housing cooling jacket directly. Due to rotor cage loss, additional cooling is required for the rotor to ensure the operating temperature of the bearings is below the thermal limit.

Both Nissan LEAF and Tesla traction motors with either series connection or parallel connection cooling jackets can be modelled using equivalent lumped parameter thermal circuits. The convective thermal resistance values of the cooling jacket are estimated from the internal flow heat transfer correlations based upon the cooling channel geometry, flow rate and coolant thermal properties.

Both the Nissan LEAF and Tesla Model S motors are radial flux machines. In comparison, the switched reluctance traction motor shown in Figure 7.22 is an axial-flux configuration [16]. The axial flux machine uses a modular stator design. There are 12 C-core modules. Each module is wound with a single phase coil being positioned over the periphery of the rotor disc. The rotor disc has 8 rotor poles. Each C-core is fixed axially by the drive end (DE) and non-drive end (NDE) endcaps, supported radially by inner support ring and trapezoidal bulges to ensure the modular stator remain rigid during operation.

Figure 7.22   *Spiral cooling jacket in the endcap of an axial flux switched reluctant machine*

Due to the axial flux topology, the cooling channels are created in the endcaps which have direct contact with the stator core rather than in the outer housing. As shown in Figure 7.22, the cooling jacket uses a double-spiral shaped design to give uniform cooling in different endcap sections. The cooling channel in the DE and NDE endcaps are connected in parallel through external connecting pipes. The water channel width and height are 16 and 8 mm, respectively, while the inner diameter of the connecting pipes varies from 11 to 15 mm. The cooling jacket design has been optimized using CFD to ensure the pressure requirement for passing 16 l/min of coolant EGW 50/50 through the cooling jacket to meet the available pump pressure. Also, CFD analysis is used to ensure the connecting pipes do not result in too much uneven flow distribution between DE and NDE. As conjugate heat transfer analysis of the cooling jacket has been performed, the convective heat transfer coefficient can be extracted from the CFD solution by using an average coolant temperature, i.e. 2,674 W/m^2 K. The heat transfer coefficient obtained by CFD is then used in an equivalent lumped parameter thermal circuit of the axial flux machine. Sensitivity analysis has been conducted using the thermal circuit [17] and found that the interface contact resistance between endcap and stator C-core has a significant impact on the winding temperature rise. To reduce the contact resistance for better cooling, thermal grease is proposed as interfacial material between the endcaps and C-cores.

The cooling medium used in electric machines with housing cooling jackets really depends on the machine application. The use of a housing cooling jacket can also be found in submersible slurry pump motors manufactured by Goodwin [18]. The Goodwin pumps are developed to have high mechanical strength, reliability and durability to operate in the most demanding environments, e.g. mining. Goodwin pumps are capable of pumping slurries up to 800 m^3/h containing abrasive solids up to 65% by weight, at up to 60 m pressure head and up to 1,000 m in distance. To cool such heavy duty motors, a pumped fluid is used as the cooling medium in relatively large channels in the motor housing. The pump is equipped with an open vane impeller at the DE to break down large particle that might ultimately lead to a blocked pump. This allows slurries to be moved around the motor housing easily.

7.7 Sleeve with flooded stator cooling

45 kW aircraft starter/generator: A case study for a machine with a sleeve and a liquid flooded stator is described in Refs [19,20]. This is a permanent magnet machine used as a starter motor/generator in an aeroengine application and operates at up to 32,000 rpm with a power of 45 kW. The losses in the stator are about 1,500 W at full power and the rotor has minimal losses of less than 100 W. An exploded view of the machine is shown in Figure 7.23 with cross-sectional views shown in Figure 7.24.

The stator is flooded with oil for cooling and a sleeve separates the oil from the rotor which runs in air. Oil enters the stator region through an annular gallery in the housing at the drive end. Oil jets from the gallery direct oil onto the end windings to provide high heat transfer coefficients. The oil then passes down square shaped

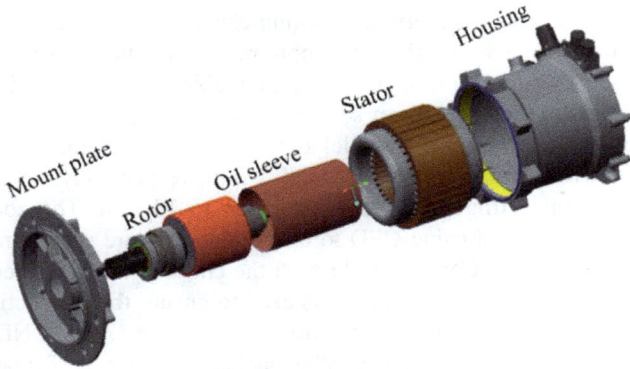

Figure 7.23   Exploded view of the sleeved, flooded stator permanent magnet
machine [19]

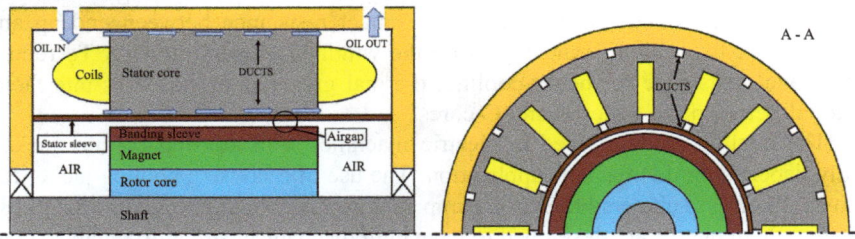

Figure 7.24   Cross-sectional views of the sleeved, flooded stator permanent
magnet machine [20]

axial ducts in the stator core. These are positioned at the outside of the laminations
bounded by the machine housing and along the gap between the stator teeth, these
are bounded by the sleeve which is positioned with a sliding fit into the inner
diameter of the stator. This reduces the size of the rotor-stator air gap but prevents
oil from coming into contact with the rotor. As oil exits the axial ducts in the stator,
deflectors direct it to impinge on the end windings at the non-drive end. This
cooling arrangement provides much more intensive cooling of the end windings
and stator than would be achieved with a water or oil jacketed cooling system.

The sleeve is made from a polymer composite. Carbon fibre composite would
be the best option as it has the greater strength, but as it is electrically conductive
there would be some eddy current losses generated. So a thicker glass fiber rein-
forced sleeve was used as this is non-conducting and there are no losses generated,
increasing the machine efficiency. Both carbon fiber and glass fiber sleeves were
manufactured and are shown in Figure 7.25, but the glass fiber sleeve was found to
be adequate in withstanding without buckling the pressure of the oil in the flooded
stator end regions. Despite the thicker glass fiber sleeve reducing the air gap, the
windage loss in the rotor at full speed was less than 100 W.

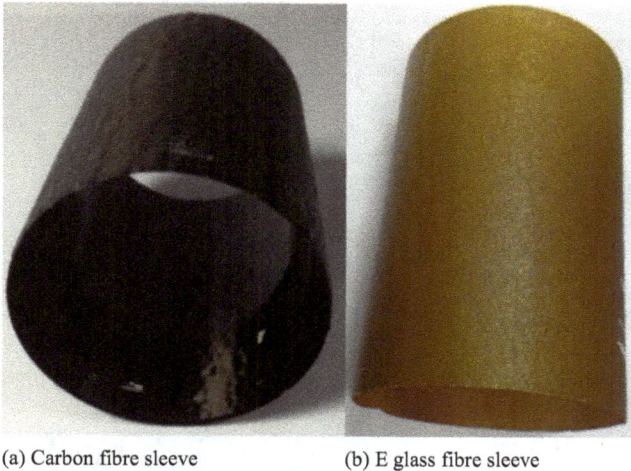

(a) Carbon fibre sleeve   (b) E glass fibre sleeve

Figure 7.25   Sleeves made for flooded stator machine [20]

The minimal loss generated in the permanent magnet rotor was dissipated through convection to the inside of the sleeve and also through the shaft to the cooled bearings.

The bearings were oil spray lubricated and this oil flow also served to provide cooling to the bearings which generated a combined loss of about 300 W.

Modelling of the thermal management system of the machine was undertaken with a combination of a lumped parameter thermal network (LPTN) based on Motor-CAD software, with the use of computational fluid dynamics (CFD) to determine the convective heat transfer coefficients for oil impingement on the end windings and in the stator ducts. These values of local heat transfer coefficient were then used in the Motor-CAD LPTN.

4 MW generator for a hybrid aircraft: Another example of a much larger machine with a flooded stator is given in Ref. [21]. This permanent magnet machine has a high power density, targeted at 18 kW/kg and is designed to produce 4 MW at 15,000 rpm. Cross-sections of the motor are shown in Figure 7.26. The permanent magnet rotor spins in air to minimize the windage loss and a glass fiber composite sleeve separates the rotor and stator. The stator is intensively cooled and is flooded with oil used as the coolant. At each end of the machine is an oil gallery with oil jets impinging on the end windings to provide high heat transfer coefficients. The end windings have a very open structure to provide a large surface area for intensive cooling on the windings. The oil then passes down cooling channels in centre of each slot between the coils. In this way, there is a minimum conduction path from the coils to the coolant thus providing direct cooling of the windings. The oil is fed into the stator from each end and leaves through a large radial duct in the centre of the stator. It then provides cooling to the outside of the stator in a cooling channels in the housing. The thermal design of the machine was undertaken using a detailed LPTN along with CFD modelling of the oil flows around the end winding.

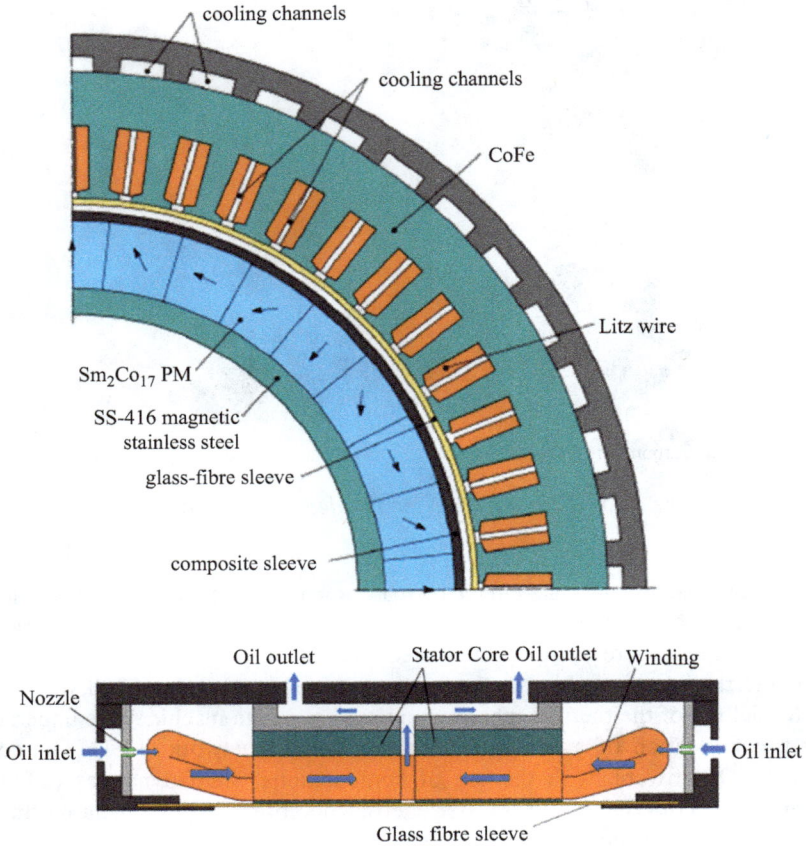

Figure 7.26   Axial and radial cross sections through a high-power density 4 MW generator for an hybrid aircraft application [21]

The cooling arrangement for the 4 MW generator represents some of the most intensive thermal management achievable in an electrical machine using a single-phase fluid. Further enhancement of heat transfer within a machine stator could also be achieved with the use of evaporative cooling in which a liquid coolant evaporates on the end windings or in ducts within the stator. Evaporative cooling gives the highest convective heat transfer coefficients achievable, but the cooling systems are inevitably more complex to design and operate. Evaporative cooling is employed in the cooling of high-power electronic chips.

7.8   Oil spray cooling

The drive for electrification in the automotive industry has resulted in high-power electric motors becoming one of the key components of the electric powertrain.

Increasing power and decreasing mass and space requirements see conventional power density limits being pushed. Hence, thermal management in the electric motors has become critical to ensure reliability and robustness in electric motors with such high-power density. By reviewing the cooling system of the existing electric motors, a housing water jacket is a common cooling solution. Since the copper loss is the major loss component, the copper loss at the end winding can be substantial. But from a heat dissipation perspective, the heat flow path of end winding is relatively long compared to active winding. As a result, the machine hot spot is usually located at the end winding. For an improved solution, the end winding can be cooled directly by an oil spray from the housing. The use of oil spray cooling can be found in the Toyota Prius hybrid vehicle [22]. Together with the improvement in stator and rotor designs, the use of oil spray cooling in the 2017 Toyota Prius could potentially increase the power density up to 36% compared to the 2009 Toyota Prius. The current density can be potentially increased up to 58% due to the effectiveness of oil spray cooling.

Nevertheless, the technical literature providing practical guidelines of how oil spray cooling should be applied to a stator with hairpin windings is limited. To fully utilize the benefits of an oil spray on electric machine cooling, more experimental investigations of oil spray cooling on an electric motor with hairpin winding is required in order to give better understanding in designing high-performance electric motors. Hairpin windings are becoming well known because they give high copper slot fill factor, reduced electrical resistance, and better thermal performance. Moreover, the gaps formed between the hairpin winding in the end space would increase the surface area that is subjected to direct oil spray cooling.

To characterize oil spray cooling performance, an experimental test rig was built as shown in Figure 7.27. Oil spray nozzles are located at 9 o'clock, 12 o'clock,

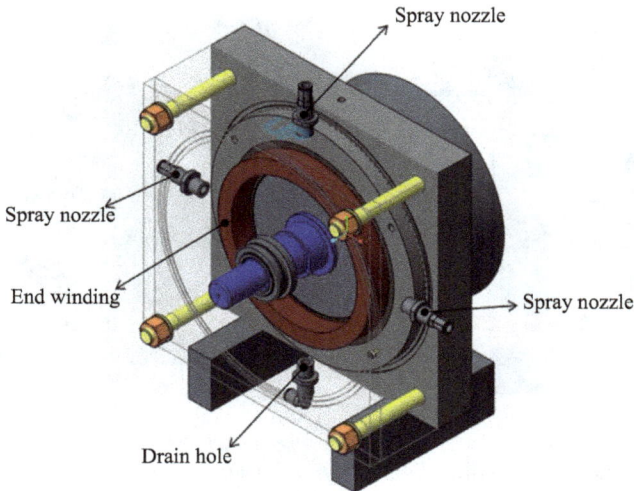

Figure 7.27   Oil spray test rig

and 3 o'clock positions, respectively. There is no nozzle at the lower part by assuming the hairpin winding at the lower part is being cooled by the falling oil film due to gravitational effects. Instead, a drain hole is at the bottom of the housing and a scavenge pump is connected to the drain hole to prevent oil from filling the end space.

Oil spray cooling is well known to give asymmetric cooling – spray coverage areas are subjected to higher heat transfer coefficients while the other end winding surfaces are cooled by a falling film of variable thickness due to gravity. Therefore, the test rig is fully instrumented with thermocouples as shown in Figure 7.28(a) in order to monitor the variation of winding temperature rise during testing. There are 23 thermocouples measuring the inner surface of the hairpin winding (T_A). As the stator consists of 72 slots, the number in Figure 7.28(b) indicates which slots are being measured. Besides the inner surface (T_A), the other surfaces are also measured such as end surface (T_B) and outer surface (T_C) of the hairpin, stator core (T_L), housing (T_{CA}), and endcap (T_{BC}). As shown in Figure 7.28(a), oil spray cooling is only applied to the drive end (DE) and there is no active cooling at the non-drive end (NDE). Besides, the temperature of end winding at NDE is also measured by thermocouples (T_{NDE}). The oil inlet and outlet temperatures are also monitored for heat removal determination, while the ambient temperature measurement is used to determine the heat dissipation to ambient from the housing. It is important to note that a static rotor is used for the tests reported here. The effects of a spinning rotor will be investigated later.

As atomization of oil involves high pressure, a high pressure pump is connected after the temperature control unit (TCU) with a built-in pump in order to increase the pumping pressure. The oil spray cooling was conducted with gas turbine lubricating oil that has a viscosity of 3 cSt at 100 °C. During testing, the oil

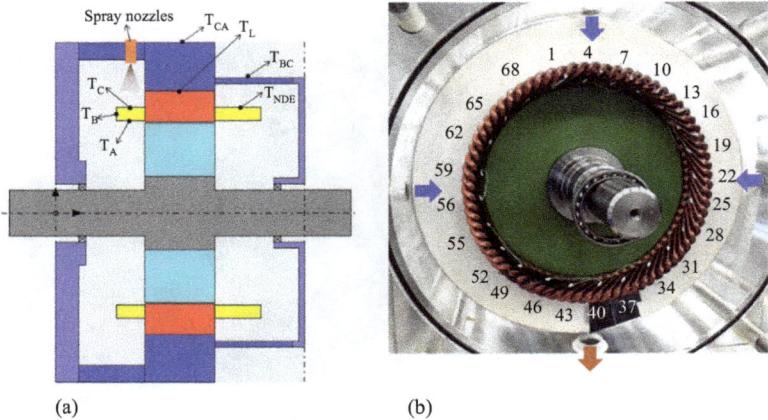

(a)   (b)

Figure 7.28   (a) A schematic diagram in axial view showing of where the thermocouples are in the test rig. (b) Thermocouple T_A measuring inner surface of hairpin winding at different slots

flow rate was varied over a range of values (1–5 l/min) and the pressure drop was measured. A manifold is used to ensure even distribution of the oil between the nozzles. To heat up the machine uniformly, all three phases of the machine are connected in series and a DC current test was performed. Supplied current and voltage are recorded during testing. In addition, the winding electrical resistances at ambient temperature and at steady-state condition are measured to determine the average winding temperature during the oil spray cooling test.

There are many different types of nozzle available in the market, but not many of them are suitable for lubricating oil. After extensive nozzle testing, two nozzle types are chosen, i.e. misting nozzle and full-cone spray nozzle. The misting nozzle creates a 90° cone-shaped mist, whereas the full-cone spray nozzle creates a 77° uniform full-cone spray. The specifications of both nozzle types are given in Table 7.8. It is important to note that each nozzle type needs to be operated under a certain range of flow rate and pressure, or the spray pattern cannot be guaranteed. Figure 7.29 shows that the pressure requirement for the misting nozzle and full-cone

Table 7.8   Misting nozzle versus full-cone spray nozzle

Nozzle type	Misting nozzle	Full-cone spray nozzle
Spray pattern	Cone-shaped mist	Uniform full cone
Nozzle diameter (mm)	1.016	0.91
Spray angle (°)	90	77
Nozzle flow rate (l/min)	0.5–3.3	0.44–1.1
Nozzle velocity (m/s)	10.3 – 67.8	11.3–28.2
Pressure (bar)	0.7–27	0.7–6

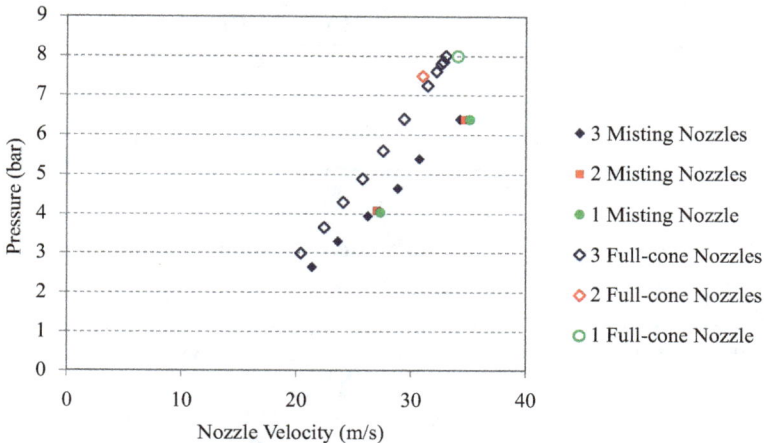

Figure 7.29   The variation of pressure loss with oil nozzle velocity for misting nozzle and full-cone nozzle

nozzle is strongly dependent on nozzle velocity (the oil velocity at the nozzle hole). The pressure loss coefficients of misting nozzle and full-cone nozzle are 1.6 and 1.2, respectively. For the same nozzle velocity, higher pressure is required for a full-cone nozzle compared to a misting nozzle.

The experimental investigation was performed by keeping the same electrical input power. The machine temperatures were recorded until the steady state condition was reached. The measured temperatures at steady state are given in Table 7.9 for the misting nozzles and Table 7.10 for the full-cone spray nozzles. A range of flow rates were tested which is distributed evenly between three nozzles through a manifold. For misting nozzles, the flow rate of 3.1–5.0 l/min results in the nozzle velocity of 21–34 m/s. Since oil spray cooling was only applied to the end winding at DE, the average temperature of the end winding at DE is much lower than that at NDE. Similar trends can be found for full-cone spray nozzles.

Besides average end winding temperatures, the temperature variation and temperature difference between the end windings at T_A based on 23 thermocouples evenly spaced around the end winding are also listed in Tables 7.9 and 7.10. For misting nozzles, by increasing the nozzle velocity from 21 to 34 m/s the temperature difference reduces from 18.9 °C to 10.4 °C. For full cone nozzles, by increasing the nozzle velocity from 20 to 33 m/s the temperature difference reduces from 32.0 °C to 20.1 °C. Hence, higher nozzle velocity reduces the temperature variation of T_A. It is important to note that the temperature variation of T_A is 3–4 °C approximately when oil spray cooling was not applied. The temperature variation

Table 7.9   *Measured temperatures after steady state with three misting nozzles over a range of flow rate*

Flow rate (l/min)	Nozzle velocity (m/s)	T_{Inlet} (°C)	T_{Outlet} (°C)	T_{DE_Avg} (°C)	T_{NDE_Avg} (°C)	T_A (°C)	ΔT for T_A (°C)
3.1	21	43.4	46.1	59.4	82.0	49.0–67.9	18.9
3.8	26	42.5	44.9	56.0	79.7	47.2–65.2	18.0
4.5	31	45.0	47.1	55.9	80.2	50.0–61.0	11.0
5.0	34	46.0	48.0	55.7	80.2	50.6–61.0	10.4

Table 7.10   *Measured temperatures after steady state with three full-cone spray nozzles over a range of flow rate*

Flow rate (l/min)	Nozzle velocity (m/s)	T_{Inlet} (°C)	T_{Outlet} (°C)	T_{DE_Avg} (°C)	T_{NDE_Avg} (°C)	T_A (°C)	ΔT for T_A (°C)
2.4	20	42.4	46.2	58.6	81.2	48.8–80.8	32.0
2.8	24	43.4	46.8	56.4	80.8	47.9–73.6	25.7
3.4	29	44.7	47.6	55.9	80.0	50.1–69.7	19.7
3.9	33	45.4	48.1	56.2	79.8	50.0–70.1	20.1

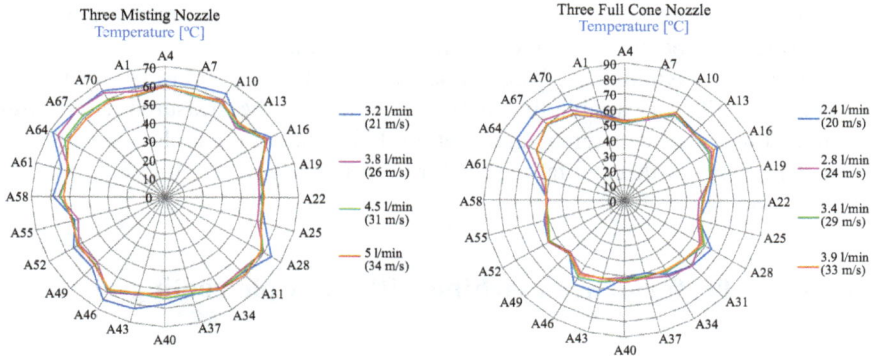

Figure 7.30 The temperature variation of T_A over different nozzle velocity for three misting nozzles (left) and full cone nozzle (right)

with three misting nozzles and full cone nozzles can be visualized easily through the radar charts as shown in Figure 7.30. Based upon the temperature variation, misting nozzles exhibit less temperature variation when compared to full cone spray nozzles. This can be explained by the hypothesis that the effect of oil mist could cool the entire end winding more evenly. Furthermore, the spray angle of the full cone nozzle is narrower than misting nozzle (i.e. 77° versus 90°). Due to the smaller spray coverage area of the full cone nozzle, the measured temperatures at 10 o'clock positions (A64, A67, and A70) and at 2 o'clock positions (A10, A13, and A16) demonstrated much higher values. On the other hand, the lower part of the end winding has similar temperatures to the upper part. This indicates that the falling oil film provides a considerable cooling effect on the end winding.

When compared to a conventional housing cooling jacket, oil spray cooling is more complicated but it has the potential to outperform machines with housing cooling jacket due to the direct contact between coolant and heat source. An extensive experimental study has been performed on oil spray cooling with hairpin windings. Two different nozzle types have been used and their performances have been compared and analysed. Based on the test data analysis, some conclusions can be made as follows:

1. The types of nozzle used for oil spray cooling play a crucial role in thermal performance as they give different spray patterns. Moreover, nozzles with a narrower spray angle tend to give higher temperature variation. Therefore, sufficient numbers of nozzles are required to ensure the entire end winding is within the spray coverage area.
2. The pressure requirement for different nozzle type varies and it strongly depends on the oil velocity at the nozzle hole. In practice, a machine designer needs to consider the available pressure in the cooling system. Together with the number of nozzles used, this will decide the nozzle velocity applied to the end winding.

3. Increased nozzle velocity not only reduces winding temperature rise but also gives less temperature variation in the end winding.
4. Due to the gravitational effect, the lower part of end winding is wetted by a falling oil film of variable thickness. By comparing the measured temperature in the lower part with temperatures at the 10 o'clock and 2 o'clock positions, it is demonstrated that the falling oil film exhibits a good cooling effect on the end winding.

7.9   High-Performance machine with multiple cooling methods

For electric motorsport applications, the electric motors need to deliver high torque over dynamic drive cycles. At the same time, the weight and volume of motors needs to be reduced. In order to design a high torque density electric motor, the machine performance is being pushed to its limit by increasing the current density. This causes challenges in thermal management of the electric motor. More advanced and effective cooling methods are required to remove the excess heat generated within the limited volume. Consequently, machine designers need to work on both electromagnetic and thermal designs together to meet the target.

For electromagnetic design, as shown in Figure 7.31, an 18 slot 16 pole synchronous permanent magnet motor with a spoke type rotor configuration is proposed to maximize the output torque through a concentrated magnetic flux topology [23]. The motor uses a tooth wound winding to ensure high compactness and a shorter end winding length for copper loss reduction. Furthermore, tooth wound windings result in a gap in the middle of the slots. The gap between the coils is used as the passage for the heat transfer fluid. For high torque density Brushless Permanent Magnet (BPM) machines, the copper loss is the major component. Moreover, the impact of AC copper losses on winding temperature rise needs to be considered as well due the effects of eddy currents at high frequency, see Figure 7.32.

To dissipate the copper losses effectively, the heat extraction fluid Paratherm LR is passed through the slots directly from one end to the other. The coolant is retained within the slots by the wedge at the slot opening as shown in Figure 7.33(a) and the flow is assumed to be evenly split between the 18 slots. In addition to direct slot cooling, the stator is also cooled by two spiral cooling channels connected in parallel in the housing. The coolant is an ethylene glycol mixture (50–50%). The benefit of the housing cooling jacket is maximized by means of minimum contact resistance between housing and stator laminations.

As the speed of the motor is up to 12,000 rpm, the Neodymium Iron Boron (NdFeB) magnets are segmented. Magnet segmentation is performed not only in the axial direction with 5 blocks per pole but also in the radial direction with 5 segments per magnet pole. Besides magnet segmentation to mitigate magnet loss, liquid cooling is used to reduce the magnet temperature to avoid demagnetization. There is a cooling duct underneath each rotor pole as shown in Figure 7.33(b). The same heat extraction fluid Paratherm LR is also used for rotor cooling.

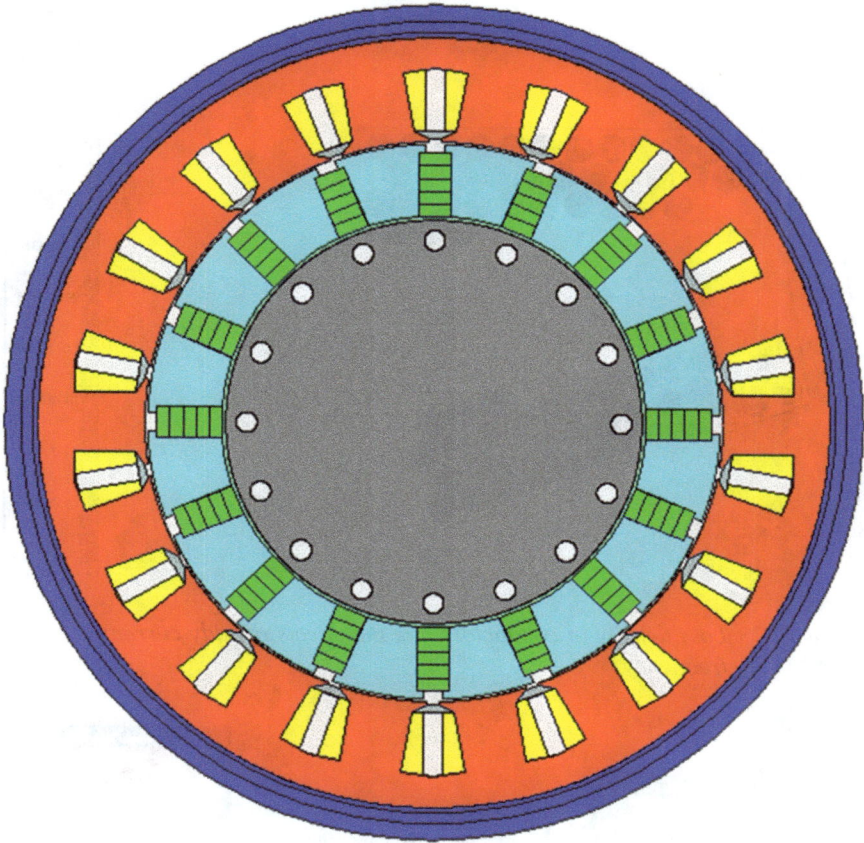

Figure 7.31 The radial cross-section of synchronous permanent magnet motor with a spoke type rotor configuration

As the electric motor is designed for electric motorsport applications, the thermal behavior over racing circuit demands need to be considered. To do that, the motor is modelled by an equivalent thermal network using Motor-CAD software. The racing circuit of Le Mans is used for this thermal study as shown in Figure 7.34(a). The machine losses over the Le Mans circuit drive cycle are computed based on 900 operating points as shown in Figure 7.34(b). The losses include winding loss with DC and AC components, stator and rotor iron loss, magnet loss, and mechanical loss. The variation of machine losses with time clearly indicate the significance of the AC winding loss component during the drive cycle especially at the high-speed points, where the AC winding loss component can be up to 4–5 times the DC winding loss component. As described above, direct slot cooling is crucial to cool the winding temperature below its thermal limit. As the copper loss is the major component and it is temperature dependent, the variation of

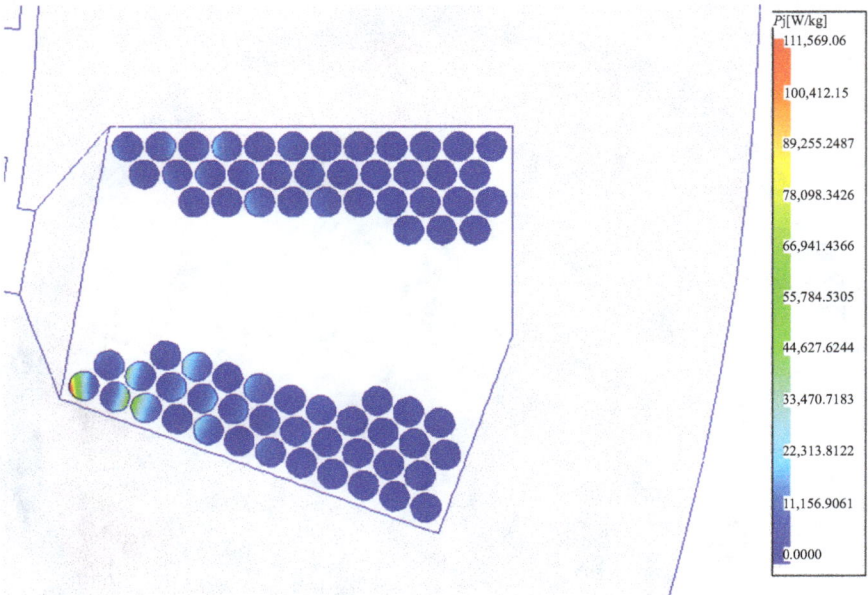

Figure 7.32   *Loss density in conductors due the effects of eddy current at high frequency*

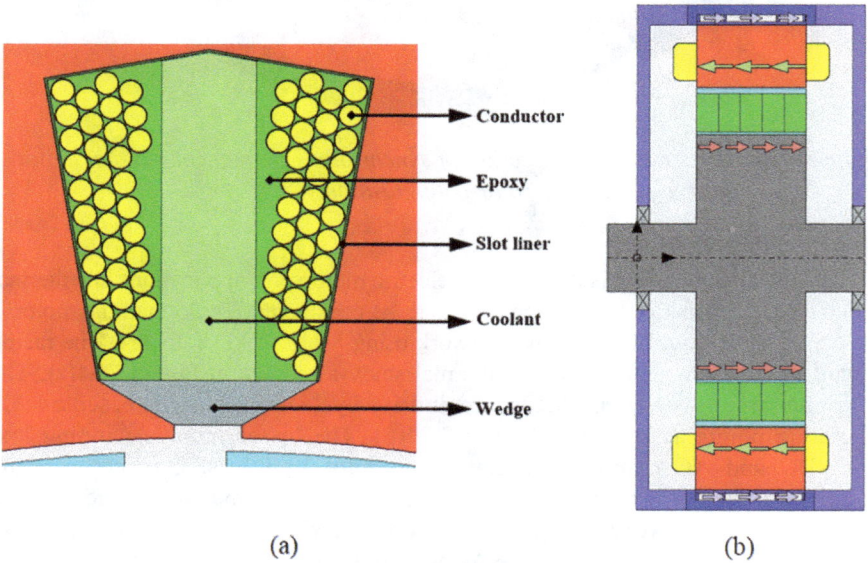

Figure 7.33   *(a) Cooling channel formed between the coils at both sides of the slot and (b) cooling systems of the electrical machine*

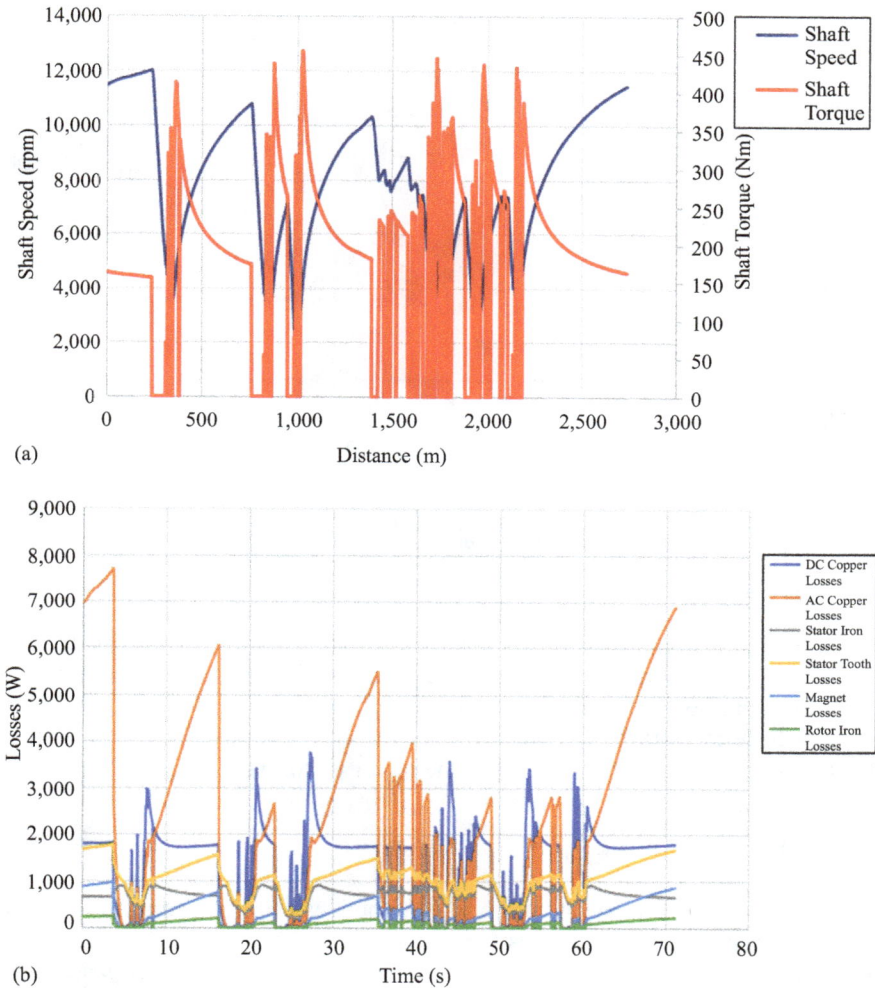

Figure 7.34 *(a) Torque and speed requirements and (b) machine losses over single lap of Le Mans circuit*

copper loss with temperature is also considered during the thermal calculation to mimic the real case scenario.

A typical race covers a distance of 85 km approximately which corresponds to 30 laps of Le Mans circuit where the track length is 2.8 km. However, the electric sports car is swapped at the half point of the race due to limitations of the batteries. As a result, the thermal performance of the electric motor with the cooling systems is simulated for 15 laps and is given in Figure 7.35. The winding reaches the steady-state temperature after 5 laps. At the steady-state condition, the maximum winding temperature varies between 140 °C and 180 °C, which is within its

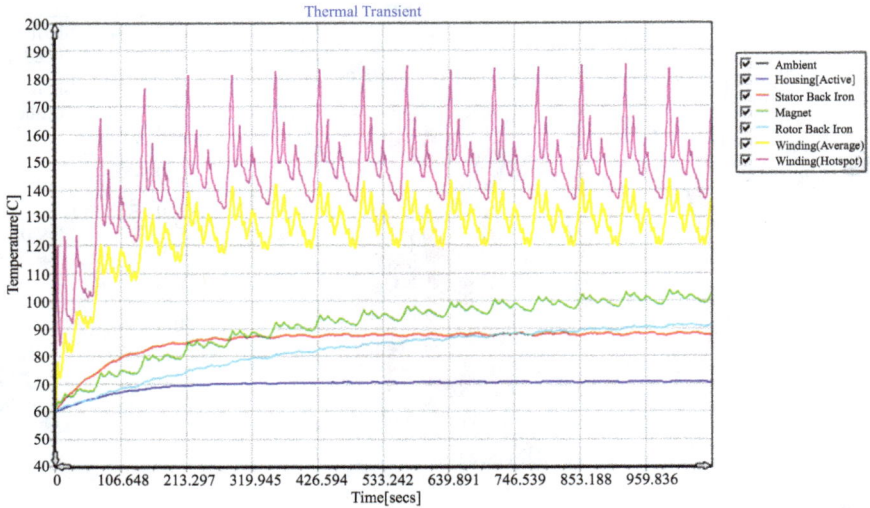

Figure 7.35   Drive cycle performance over 15 laps of Le Mans circuit

insulation thermal limit class H. The rotor takes more laps to reach the steady state, but the magnet temperature is well below its thermal limit of 150 °C over the race. To design a high-performance electric motor for motorsport applications, designers not only have to work on the electromagnetic design but also have to pay great attention to the thermal design to meet the motor performance target.

References

[1]   Hollstein W. 'Epoxy: solutions for automotives'. *Epoxy Resins in Electrical Engineering Symposium*, Ostfildern, Germany; 2019.

[2]   Huntsman Corporation. *Reliable Chemistry for E-motor: Araldite® Heat-Conductive Encapsulants and Trickle and Dipping Impregnation Epoxy Resins for E-motors Improve Heat Dissipation and Extend Lifetime* [online]; 2021. Available from https://www.huntsman-transportation.com/auto-motive-emobility/e-motor.html [Accessed 3 November 2021].

[3]   LORD Corporation. *Technical Data: CoolThermTM EP-340 Epoxy Resin*; 2017.

[4]   Staton D. A., Hawkins, D., and Popescu, M. 'Thermal behaviour of electrical motors – an analytical approach'. *INDUCTICA Technical Conference*; CWIEME, Berlin, Germany; 2009, pp. 1–8.

[5]   Sawata T. and Staton D. 'Thermal modeling of a short-duty motor'. *IECON 2011 – 37th Annual Conference of the IEEE Industrial Electronics Society*; Melbourne, VIC, Australia, November 2011. New York, NY: IEEE; 2012, pp. 2054–2059.

[6] Staton D., Boglietti A., and Cavagnino A. 'Solving the more difficult aspects of electric motor thermal analysis in small and medium size industrial induction motors'. *IEEE Trans. Energy Convers.* 2005;20(3):620–628.

[7] Boglietti A., Cavagnino A., Staton D. 'Determination of critical parameters in electrical machine thermal models'. *IEEE Trans. Ind. Appl.* 2008;44(4): 1150–1159.

[8] Nategh S., Zhang H., Member S., *et al.* 'Transient thermal modeling and analysis of railway traction motors'. *IEEE Trans. Ind. Electron.* 2019;66(1): 79–89.

[9] Akayshee Q. and Williams K. '1500 hp AC drive motor design for oil field drilling operations'. *Offshore Technology Conference*; 2003.

[10] ATB Laurence Scott. *High Voltage Induction Motors* [Online]; 2015. Available from http://www.laurence-scott.com/downloads/ [Accessed 22 September 2018].

[11] Brush. *Combined (Hydrogen & Water) Cooled 2-Pole Turbogenerators*, Brush Turbogenerators; 2008.

[12] Zytek. *Zytek 170 kW 460 Nm Electric Traction Motor* [Online]; 2018. Available from http://www.zytekautomotive.co.uk/products/electric-engines/170kw/ [Accessed 18 November 2018].

[13] Bennion K., Cousineau E., Feng X., King C., and Moreno G. 'Electric motor thermal management R&D'. Presented at IEEE Power & Energy Society General Meeting, Denver, CO; 2015.

[14] Sato Y., Ishikawa S., Okubo T., Abe M., and Tamai K. 'Development of high response motor and inverter system for the Nissan LEAF electric vehicle'. *SAE 2011 World Congress & Exhibition*; SAE Technical Paper 2011-01-0350; 2011, pp. 1–8.

[15] Kim K. 'Driving characteristic analysis of traction motors for electric vehicle by using FEM', *Ansys Electron. Expo*; 2014, pp. 1–39.

[16] Europe Commission CORDIS. *Final Report Summary – VENUS (Switched/ Synchronous Reluctance Magnet-free Motors for Electric Vehicles)*, Seventh Framework Programme. Project ID: 605429; 2017.

[17] Chong Y. C., Staton D., Egaña I., de Argandoña R., and Egea A. 'Thermal design of a magnet-free axial-flux switch reluctance motor for automotive applications'. *2016 Eleventh International Conference on Ecological Vehicles and Renewable Energies (EVER)*, Monte Carlo, Monaco, April 2016. New York, NY: IEEE; 2016, pp. 1–8.

[18] Goodwin India. *Submersible Slurry Pump* [online]; 2017. Available from https://www.goodwinindia.in [Accessed 19 November 2018].

[19] Xu Z., La Rocca A., Arumugam P., *et al.*, 'A semi-flooded cooling for a high speed machine: concept, design and practice of an oil sleeve'. *IECON 2017 – 43rd Annual Conference of the IEEE Industrial Electronics Society*, Beijing, China; November 2017. New York, NY: IEEE; 2017, pp. 8557–8562.

[20] La Rocca A., Xu Z., Arumugam P., *et al.* 'Thermal management of a high speed permanent magnet machines for an aeroengine'. *2016 XXII International Conference on Electrical Machines (ICEM)*, Lausanne, Switzerland; September 2016. New York, NY: IEEE; 2016, pp. 1–6.

[21]   Golovanov D., Gerada D., Sala G., *et al.* '4MW Class High Power Density Generator for Future Hybrid-Electric Aircraft'. *IEEE Trans. Transp. Electrification*. 2021;7(4):2952–2964.

[22]   Sano S., Yashiro T., Takizawa K., and Mizutani T. 'Development of new motor for compact-class hybrid vehicles'. *World Electr. Veh. J.* 2016;8 (2):443–449.

[23]   Volpe G., Chong Y. C., Staton D. A., and Popescu M. 'Thermal management of a racing E-machine'. *2018 XIII International Conference on Electrical Machines (ICEM)*, Alexandroupoli, Greece, September 2018. New York, NY: IEEE; 2018, pp. 2689–2694.

Index